Principles of
Clinical Genetics

Principles of
Clinical Genetics

Yogesh Ashok Sontakke
MBBS MD (Anatomy)
Assistant Professor
Department of Anatomy
Jawaharlal Institute of Postgraduate Medical Education and Research (JIPMER)
(An Institute of National Importance under the Ministry of Health and
Family Welfare, Government of India)
Puducherry, India

The Health Sciences Publisher
New Delhi | London | Panama

Jaypee Brothers Medical Publishers (P) Ltd

Headquarters
Jaypee Brothers Medical Publishers (P) Ltd
4838/24, Ansari Road, Daryaganj
New Delhi 110 002, India
Phone: +91-11-43574357
Fax: +91-11-43574314
Email: jaypee@jaypeebrothers.com

Overseas Offices

J.P. Medical Ltd
83 Victoria Street, London
SW1H 0HW (UK)
Phone: +44-20 3170 8910
Fax: +44 (0)20 3008 6180
Email: info@jpmedpub.com

Jaypee-Highlights Medical Publishers Inc
City of Knowledge, Bld. 235, 2nd Floor, Clayton
Panama City, Panama
Phone: +1 507-301-0496
Fax: +1 507-301-0499
Email: cservice@jphmedical.com

Jaypee Brothers Medical Publishers (P) Ltd
17/1-B, Babar Road, Block-B, Shaymali
Mohammadpur, Dhaka-1207
Bangladesh
Mobile: +08801912003485
Email: jaypeedhaka@gmail.com

Jaypee Brothers Medical Publishers (P) Ltd
Bhotahity, Kathmandu
Nepal
Phone: +977-9741283608
Email: kathmandu@jaypeebrothers.com

Website: www.jaypeebrothers.com
Website: www.jaypeedigital.com

© 2018, Jaypee Brothers Medical Publishers

The views and opinions expressed in this book are solely those of the original contributor(s)/author(s) and do not necessarily represent those of editor(s) of the book.

All rights reserved. No part of this publication may be reproduced, stored or transmitted in any form or by any means, electronic, mechanical, photocopying, recording or otherwise, without the prior permission in writing of the publishers.

All brand names and product names used in this book are trade names, service marks, trademarks or registered trademarks of their respective owners. The publisher is not associated with any product or vendor mentioned in this book.

Medical knowledge and practice change constantly. This book is designed to provide accurate, authoritative information about the subject matter in question. However, readers are advised to check the most current information available on procedures included and check information from the manufacturer of each product to be administered, to verify the recommended dose, formula, method and duration of administration, adverse effects and contraindications. It is the responsibility of the practitioner to take all appropriate safety precautions. Neither the publisher nor the author(s)/editor(s) assume any liability for any injury and/or damage to persons or property arising from or related to use of material in this book.

This book is sold on the understanding that the publisher is not engaged in providing professional medical services. If such advice or services are required, the services of a competent medical professional should be sought.

Every effort has been made where necessary to contact holders of copyright to obtain permission to reproduce copyright material. If any have been inadvertently overlooked, the publisher will be pleased to make the necessary arrangements at the first opportunity. The **CD/DVD-ROM** (if any) provided in the sealed envelope with this book is complimentary and free of cost. **Not meant for sale.**

Inquiries for bulk sales may be solicited at: jaypee@jaypeebrothers.com

Principles of Clinical Genetics

First Edition : **2018**
ISBN: 978-93-5270-188-9
Printed at Nutech Print Services - India

Dedicated to

*My wife **Dr Anindita**
and daughter **Aripra**
for their continuous love…*

Preface

Clinical genetics is a rapidly advancing branch of medicine. Understanding the basic principles of clinical genetics is essential for the diagnosis, management and prevention of various disorders.

Due to the burden of medical subject content, students face difficulty in focusing attention on genetics. Hence, in the present book, facts about clinical genetics are given in a concise and simple manner. Most of the chapters are provided with the clinical problems and their solutions to make the subject interesting. This also helps to convey the message of clinical utility of genetics.

For preparation of postgraduate entrance examination, it is difficult to read a new book on genetics. Thus, in this book, National Eligibility Cum Entrance Test (NEET) markings are given at the information for entrance examination and multiple choice questions.

Viva markings are also given to help the students for oral examinations. Most of the sections are provided with questions to understand the importance of the topic. To make it easy, text is supported with flow charts. Annexures cover examination-related topics. The book follows the course contents of All India Institute of Medical Sciences (AIIMS), Jawaharlal Institute of Postgraduate Medical Education and Research (JIPMER), Maharashtra University of Health Sciences (MUHS)-Nasik and many other universities and medical institutions.

I hope that the book will prove to be helpful as a teaching and learning resource for graduate and postgraduate students of medicine, biotechnology and allied health sciences.

Yogesh Ashok Sontakke

Acknowledgments

First and foremost, I wish to thank, Professor (Dr) Subhash Chandra Parija, Director, Jawaharlal Institute of Postgraduate Medical Education and Research (JIPMER), for his inspiration and generous support. I am greatly indebted to Dr Parkash Chand, Professor (Senior Scale) in Anatomy, JIPMER for his encouragement, invaluable support. I extend my sincere gratitude to Dr K Aravindhan, Additional Professor and Head, Department of Anatomy, JIPMER, for his motivation and suggestions.

I am grateful to Shri Jitendar P Vij (Group Chairman) and Mr Ankit Vij (Group President) who are heading the experienced team of M/s Jaypee Brothers Medical Publishers (P) Ltd, New Delhi, India, for giving the shape to this book. Special thanks to Ms Ritu Sharma (Director-Content Strategy), for continuous support.

I acknowledge the supports from Drs M Sivakumar, Suma HY, Sarasu J (Additional Professors), and Dr V Gladwin, Associate Professor (Anatomy), JIPMER. Special thanks to Dr Dharmaraj Tamgire (Assistant Professor), for his encouragement and constructive criticism. I would also like to acknowledge Drs Raveendranath V, Suman Verma, Rajasekhar SSSN, Sulochana Sakthivel [Assistant Professors (Anatomy)], JIPMER, for their continuous unconditional support. I also acknowledge Drs G Vidya, Thuslima M, Vijaykishan Bheemavarapu, Sujithaa N (Senior Residents), for their support. I also acknowledge Drs Anitha B, Kiran K, Shanthini S, Phoebe Johnson, Challa Ravi, Rajeev Panwar, G Dhivya Lakshmi, Saleena N Ali, Tom J Nallikuzhy, Arun Prasad and Sabin Malik (Junior Residents) for their help.

I thank my wife Dr Anindita, for editing and proofreading. Her devotion, unconditional love and support, sense of humor, patience, optimism and advice were more valuable than one could ever imagine. I appreciate my little girl Aripra, for her unconditional love that inspired me to write the book. My sincere regards to my parents and parents-in-law for their moral support. I thank the Almighty for giving me the strength and patience to work.

Contents

Chapter 1 **Introduction**   1
Points to Remember *1*
Branches of Genetics *2*

Chapter 2 **Chromosomes**   3
Morphology of Chromosomes *3*
Structure of Chromosome *4*
Euchromatin and Heterochromatin *5*
Chemical Composition of Chromosomes *7*
Structural Classification of Chromosomes *8*

Chapter 3 **Cytogenetics**   10
Karyotyping *10*
Fluorescent *in Situ* Hybridization *13*

Chapter 4 **Chromosomal Aberration**   15
Classification of Chromosomal Aberrations *15*
Structural Chromosomal Aberrations *15*
Genomic Imprinting *17*
Factors Causing Chromosomal Aberrations *20*
Numerical Chromosomal Abnormalities *20*
Down's Syndrome *22*
Turner Syndrome *23*
Klinefelter Syndrome *25*
Patau Syndrome *26*
Edward Syndrome *27*
Hermaphroditism *27*

Chapter 5 **Structure of DNA and RNA**   29
Deoxyribonucleic Acid *29*
Ribonucleic Acid *31*
Differences between DNA and RNA *33*
Satellite DNA *33*
Mitochondrial DNA *33*
Transcription and Translation *34*
Gene *34*

Mutation 38
Gene Mapping 39
Gene Bank 41

Chapter 6 Laws of Inheritance   42
Characters Studied by Mendel 43
Mendel's Laws 43
Biological Significance of Mendel's Laws 46

Chapter 7 Patterns of Inheritance   47
Pedigree 48
Single Gene Inheritance 49
Polygenic Inheritance (Multifunctional Inheritance) 56
Mitochondrial Inheritance 58

Chapter 8 Inborn Errors of Metabolism   60
Pathogenesis of Inborn Errors of Metabolism 60
Phenylketonuria 62
Solution for Clinical Case 63

Chapter 9 Dermatoglyphics   64
Anatomy of Fingerprint 65
Patterns in Dermatoglyphics 65
Galton Classification of Finger Patterns 66
Significance of Dermatoglyphics 68

Chapter 10 Cancer Genetics   69
Cancer-causing Genes 69
TP53 Gene 71

Chapter 11 Prenatal Diagnosis   73
Methods of Prenatal Diagnosis 73
Amniocentesis 75
Chorionic Villus Sampling 77
Percutaneous Umbilical Blood Sampling 78

Chapter 12 Gene Therapy   80
History of Gene Therapy 80
Principles of Gene Therapy 80
Approaches for Gene Therapy 81

Gene Transfer Techniques *82*
Physical Methods for Gene Delivery *83*
Biological Gene Transfer *85*
Chemical Methods of Gene Transfer *86*
Medical Conditions Treated with Gene Therapy *87*
Disadvantages and Hurdles in Gene Therapy *87*

Chapter 13 **Stem Cell Therapy** 88
Classification of Stem Cells *89*
Embryonic Stem Cells *89*
Adult Stem Cells *91*
Induced Pluripotent Stem Cells *92*
Applications of Stem Cell Therapy *92*
Cord Blood Bank *93*

Chapter 14 **Genetic Counseling** 95
Steps of Genetic Counseling *96*

ANNEXURES

Annexure 1 **Polymerase Chain Reaction** 99
Required Material *99*
Procedure *100*
Applications of Polymerase Chain Reaction *100*

Annexure 2 **Recombinant DNA Technology** 104
Procedure *104*
Applications of Recombinant DNA Technology *104*

Annexure 3 **DNA Fingerprinting or Profiling** 106
Protocol *106*

Annexure 4 **Developmental Genetics** 107
HOX (Homeobox) Genes *107*
Paired Box (*PAX*) Gene *107*

Annexure 5 **SRY Gene** 109
Role of *SRY* Gene *109*
Mutation of *SRY* Gene *109*

| Annexure 6 | **Hydatidiform Mole** | 110 |

Mode of Formation *110*
Genomic Basis *110*

| Annexure 7 | **Blood Group Genetics** | 111 |

Multiple Genes *111*
ABO Blood Group System *111*
Rh-Factor *112*

| Annexure 8 | **Immunogenetics** | 114 |

Immunity *114*
Antigen *114*
Antibody *114*
Types of Grafts *114*
Major Histocompatibility Complex or Human Leukocyte Antigen System *115*
Human Leukocyte Antigen Complex *116*

| Annexure 9 | **Twins** | 117 |

Incidence *117*
Types of Twins *117*

| Annexure 10 | **Cloning** | 120 |

Methods of Cloning *120*

Index   *121*

CHAPTER 1

Introduction

POINTS TO REMEMBER

- The term genetics was first coined by William Bateson (1906). (Greek: gene = to grow into).
- Genetics is defined as the branch of the biology that deals with heredity and variation.
- Heredity is the inheritance of character traits from parents to offspring.
- Variation indicates the differences among the offspring and also from parents.
- Genetics involves the study of the mechanisms of heredity by which characters are transmitted or inherited from generation to generation.
- Some of the important historical milestones of the genetics are listed in **Table 1**.

Table 1: History of human genetics

Year	Milestone
1865–66	Gregor Mendel proposed laws of inheritance (Father of Genetics)
1875	*Galton:* Polygenic inheritance, the effect of environment on heredity in twins
1900	*Landsteiner:* Discovery of ABO blood group
1902	*Garrod:* Reported alkaptonuria as an example of Mendelian inheritance in human
1952	*Gerty Cori and Carl Cori:* Glucose-6-phosphate deficiency → glycogen storage disease
1953	*Jervis:* Phenylalanine hydroxylase deficiency → Phenylketonuria (PKV)
1956	*Tjio and Levan:* Human cells have 46 chromosomes
1959	*Lejeune:* Chromosomal aberration → Down syndrome
1959	*Ford and Jacobs:* The role of Y chromosome in sex determination
1961	*Watson, Crick, Wilkins:* Structure of DNA
1968	*Har Gobind Khorana, Nirenberg, Holley:* Genetic code and its role in protein synthesis
1976	*Smith, Nathans, Arber:* Use of restriction enzymes in DNA research
1983	*Barbara McClintock:* Jumping genes
1993	*Richard Roberts, Phillip Sharp:* Studies on genes, DNA and interns
2002	*Brenner, Horvitz, Sulston:* Genetic regulation of organ development and programmed cell death
2006	*Fire and Mello:* Gene silencing by double-stranded RNA
14th April, 2003	Compilation of Human Genome Project
2016	Kate Rubins (NASA)—a genome is sequenced in space at International Space Station

Abbreviations: DNA, deoxyribonucleic acid; RNA, ribonucleic acid; NASA, National Aeronautics and Space Administration

BRANCHES OF GENETICS

Genetics is divided into the following branches:
- *Cytogenetics:* Study of chromosomes.
- *Molecular genetics:* Study of the chemical structure of the genetic material and its role.
- *Biochemical genetics:* Biochemical study of genetic material including inborn errors of metabolism.
- *Cancer genetics:* Study of genes concerned with cell cycle and cancers.
- *Developmental genetics:* Study of genes concerned with development.
- *Immunogenetics:* Study of causative genetic aspects of immunity.
- *Clinical genetics:* Study of causative genetic factors in a clinical setup.
- *Population genetics:* Study of laws of inheritance in the human population.
- *Eugenics:* Study of applications of principles of heredity for improvement of species.

CHAPTER 2

Chromosomes

CLINICAL CASE

A physician made a call to a genetic center for obtaining the information regarding a 19-year-old female patient with primary amenorrhea with the absence of Barr body on buccal smear. The physician inquired about the reasons for the absence of Barr body. How will you respond to his query?

INTRODUCTION

- Chromosomes in nondiving cells (interphase) are not visible under microscope because they are present in the form of long thread-like material called **chromatin**.
- During cell division, chromatin condenses to form chromosomes (visible under light microscope).
- Chromatin consists of deoxyribonucleic acid (DNA) and present in the nuclei of cells **(Box 1)**.
- Chromosomes = chrom—color and soma—body in Greek.
- Waldyer coined the term 'chromosome' (1888).
- Robert Feulgen (1924) showed that the chromosomes contain DNA.
- Kaufmann (1948) described the morphology of chromosomes.

Box 1: DNA to chromosome

DNA in nucleus → chromatin in interphase → chromosomes during cell division.

MORPHOLOGY OF CHROMOSOMES

Morphology of the chromosomes can be best studied during metaphase of cell division. In other phases (prophase, anaphase and telophase), chromosomes are either not completely condensed or over condensed.

Number

- The number of chromosomes in the given species is constant.
- *Diploid (2n) number:* The number of chromosomes in each of the somatic cells (46).
- *Haploid (n) number:* The number of chromosomes in each of the gametes (sperm or ovum). They are 23 in human.

Shape
- Shape of the chromosome varies according to the phase of cell division.
    - *In interphase:* Long, thin, thread-like (chromatin)
    - *Prophase:* Long, condensed chromosomes
    - *Metaphase:* Thick, condensed filamentous
    - *Anaphase:* Short, divided at centromere.

STRUCTURE OF CHROMOSOME (FIG. 1)

Q. Describe the structure of chromosome.

Chromatid *(Viva)*
- At mitotic metaphase, each chromosome shows **two symmetrical** structures called **sister chromatids**.
- Both chromatids are attached to each other at the **centromere**.
- Sister chromatids get separated at centromere in anaphase and migrate to the opposite poles.

Centromere or Kinetochore

Q. Define centromere.
- *Definition:* Centromere is the region of the chromosome to which sister chromatids are attached.
- **Primary constriction:** Chromosome forms a thin segment of the chromosome called primary constriction.
- Centromere consists of repetitive DNA sequences that bound to specific proteins and thus, it forms disc-shaped structure called **kinetochore**.
- During mitosis, microtubules of mitotic spindle get attached to the kinetochore and provide the force for chromosomal movement during anaphase.
- Human chromosomes are **monocentric** as they contain only one centromere.
- Due to X-ray exposure, chromosome breaks and may form abnormal chromosomes → acentric (no centromere) or dicentric (two centromeres).

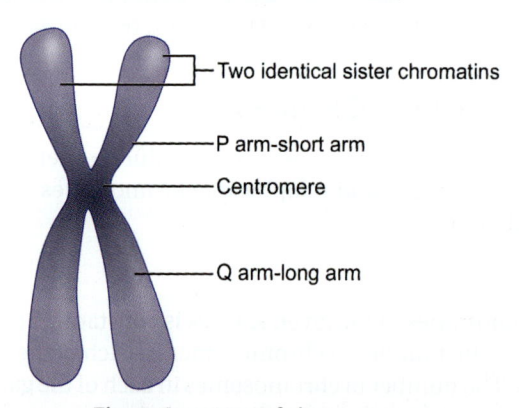

Fig. 1: Structure of chromosome

Telemere
- **Free ends** of sister chromatids are known as telomeres.
- Telemeres prevent fusion of chromosomes with each other.
- Deletion of telomere makes the end sticky.
- Telomerase is the enzyme responsible for telomere synthesis and maintenance of length of the telomeres. Telomerase is RNA-dependent DNA polymerase. *(NEET)*
- Telomerase activity is present in germ cells and stem cells (hematopoietic cells) but absent in somatic cells. Increased telomerase activity favours cancer cells. *(NEET)*

Secondary Constriction
- A narrowed portion of chromosome other than centromere is called as secondary constriction.
- Secondary constriction helps in the identification of chromosomes.

Nuclear Organisers
- Nuclear organiser region (NOR) are the regions of secondary constrictions that contain genes for coding ribosomal RNA.
- NOR forms *nucleolus*.
- In human, chromosomes 13, 14, 15, 21 and 22 (all acrocentric chromosome) have NOR.

Satellite
- Satellite is round, elongated knob-like appendage of chromosomes.
- Only acrocentric chromosomes (13, 14, 15, 21 and 22) show satellite bodies.
- Satellites are connected with rest of the chromosomes by a thin filament of chromatin.

Note
- **Microsatellite** sequence is a short sequence of DNA *repeats* (2–6 base pairs). *(NEET)*
- Centromere and telomere have highly repetitive DNA sequences. *(NEET)*

EUCHROMATIN AND HETEROCHROMATIN

Depending on the shade of staining, chromatin is classified as lightly stained euchromatin and darkly stained heterochromatin **(Table 1 and Fig. 2)**.

Euchromatin
Partially condensed portion of chromatin stains lightly and it is called **euchromatin**.

Heterochromatin
Q. **Write a short note on heterochromatin.**
- The **condensed** part of the chromatin stains **darkly** and it is called heterochromatin.
- It has high content of repetitive DNA sequences and very few structural genes.
- It is late replicating and nontranscribed part of chromatin material.

Types of Heterochromatin
- Heterochromatin is classified into the following two types: constitutive and facultative.

Table 1: Differences between euchromatin and heterochromatin

Euchromatin	Heterochromatin
Lightly stained chromatin	Darkly stained chromatin
Less coiled chromatin *(NEET)*	Condensed, highly coiled chromatin
Active chromatin *(NEET)*	Inactive chromatin
Expresses during interphase	Do not express during interphase
Contains structural genes	Contains repetitive DNA sequences
Mutations occur frequently in euchromatin	Mutations occur less frequently in heterochromatin
Contains guanine and cytosine bases predominantly	Contains adenine and thymine bases predominantly

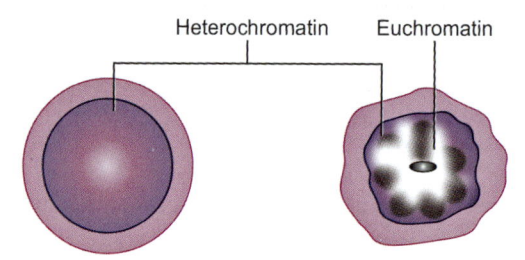

Fig. 2: Heterochromatin and euchromatin

Constitutive Chromatin

- *Definition*: The part of the heterochromatin that remains **inactive** throughout the cell cycle or life of an organism is called constitutive heterochromatin.
- *Location*: Around centromere, telomeres and C-bands of chromosomes. *(NEET)*
- In human, chromosomes 1, 9, 16, 19 and Y show significant constitutive heterochromatin. *(NEET)*
- *Method for detection*: C-banding *(NEET)*
- *Function*: During cell division, constitutive heterochromatin is necessary for proper segregation of sister chromatids and centromere.
- *Clinical aspects:*
  - Robert syndrome (Hypomelia-hypotrichosis-facial hemangioma) is an autosomal recessive disorder caused by mutation at *ESCO2* gene of 8th chromosome. *(NEET)*
  - Immunodeficiency, centromere instability and facial anomalies (ICF) syndrome is a autosomal recessive disorder caused by mutation of DNA-methyltransferase-3b gene on chromosome 20.

Facultative Heterochromatin

- *Definition*: The part of heterochromatin that remains **active** at certain phase of cell cycle or life of an organism is called facultative heterochromatin.
- *Example*: One X chromosome remains active throughout the life but another X chromosome is active only during embryogenesis and female reproductive organ formation. Later, second X chromosome becomes inactive and forms facultative heterochromatin.
- Inactive X chromosome form Barr body **(Box 2)**.

Box 2: Barr body or sex chromatin

Q. Write a short note on Barr body.
- Barr body was detected by Murray Barr (1948).
- *Definition:* Nucleus of diploid cells shows presence of dense dark-staining spot at the periphery of the nucleus that represent inactive X chromosome. This darkly stained zone is called as Barr body.
- In 3% of **neutrophil**, inactive X chromosome appears as **drumstick**. *(NEET)*
- Number of Barr bodies = Number of X chromosomes − 1 **(Fig. 3) (Table 2)**.

Lyon Hypothesis

Q. Write a short note on Lyon hypothesis.
- It states that in female cells, only one X chromosome is active. Another X chromosome remains inactive. The process of inactivation is called as **lyonisation**.
- *Mechanism of inactivation:* **Methylation** of nucleotide bases in X chromosome begins at 15–16 days of intrauterine life that makes it inactive.

Table 2: Number of Barr bodies

Condition	Karyotype	Number of Barr bodies
Normal male	46,XY	Absent
Normal female	46,XX	1
Turner syndrome	45,X	Absent
Klinefelter syndrome	47,XXY	1
XXX syndrome	47,XXX	2

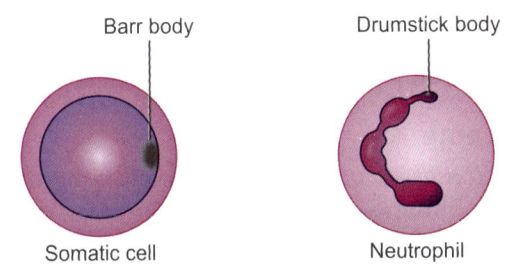

Fig. 3: Barr body

CHEMICAL COMPOSITION OF CHROMOSOMES

- Chemical composition of the chromosome is given in **Flow chart 1**.
- Histone proteins are rich in basic amino acids and help in packaging of the DNA in chromosomes. *(NEET)*
- H1 histone protein is a linker molecule, hence does not appear in core histone proteins. *(NEET)*
- Nucleosome consists of DNA wrapped around the core histone protein. *(NEET)*
- Each histone protein is wrapped by 146 base pairs of DNA. *(NEET)*

Flow chart 1: Chemical composition of chromosome

```
                    Chemical composition of chromosome
                                    │
        ┌───────────────────────────┼───────────────────────────┐
        ▼                           ▼                           ▼
       DNA                   Histone proteins              Nonhistone proteins
                                    │                           │
                                    ▼                           ▼
                         H1, H2A, H2B, H3 and H4    Actine, myosin, α- and β-tubulins
                         • Important for repression • Helps in chromosome condensation,
                           of genes                   DNA replication, and transcription
                         • Maintain integrity of
                           chromosomes
```

Abbreviation: DNA, deoxyribonucleic acid

STRUCTURAL CLASSIFICATION OF CHROMOSOMES

Q. Classify the chromosome (Box 3).
- Each chromosome has two arms—short arm or p-arm (petit = small) and long arm or q-arm (*q* is next alphabet to *p*). *(NEET)*
- Both the arms are connected with each other by centromere.
- According to the position of centromere, chromosomes are classified as metacentric, submetacentric, acrocentric and telocentric chromosomes **(Fig. 4)**.

Metacentric Chromosome
Chromosome with central centromere (***p-arm length = q-arm length***) is called metacentric chromosome, for example, chromosomes 1 and 3. *(NEET)*

Submetacentric Chromosome
Chromosome with centromere near the center (***p-arm length < q-arm length***) is called submetacentric chromosome, for example, chromosomes 4 and 5.

Acrocentric Chromosome
Chromosome with centromere at one end with satellite bodies is called acrocentric chromosomes, for example, chromosomes 13, 14 and 15.

Telocentric Chromosome
- Chromosome with centromere at the one end (have **only one arm**) is called telocentric chromosome.
- In human, telocentric chromosomes are absent.

Box 3: Facts about chromosomes
- Chromosomes are also classified as sex chromosomes (X and Y chromosomes) and autosomes (1–22 chromosomes).
- Chromosomal nomenclature is based on centromeric position. *(NEET)*

| Metacentric | Submetacentric | Acrocentric | Telocentric |
| p-arm = q-arm | p-arm = q-arm | | |

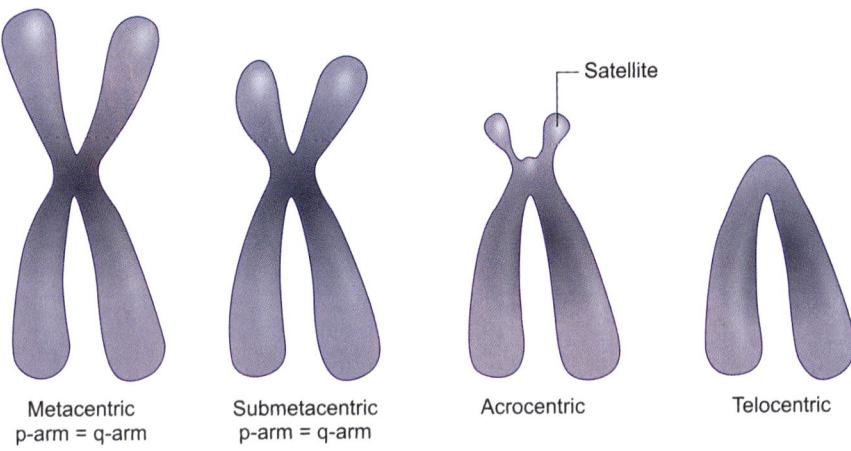

Fig. 4: Types of chromosomes

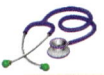

SOLUTION FOR CLINICAL CASE

The nucleus of diploid cells shows the presence of dense dark-staining spot at the periphery of the nucleus that represents inactive X chromosome. This darkly stained zone is called as Barr body. Number of Barr bodies = Number of X chromosomes – 1. Thus, normal female with 46,XX karyotype shows the presence of single Barr body. In the present case, as the female patient with the absence of Barr body and primary amenorrhea indicates a case of Turner syndrome (45,X). For confirmation of the condition, karyotyping should be advised.

CHAPTER 3

Cytogenetics

CLINICAL CASE

A 65-year-old male patient with generalised lymphadenopathy, increased total leukocyte count (20×10^9/L), lymphocytosis with smudge cells (fragile leukemic cells), hypercellular bone marrow is referred for karyotyping. The patient is a suspected case of chronic myeloid leukemia. What method will you suggest for cytogenetic analysis?

INTRODUCTION

- Cytogenetics is a branch of genetics that studies the number and structure of the cells.
- Cytogenetic study involves following methods:
  - Karyotyping
  - Fluorescent *in situ* hybridization (FISH)
  - Buccal smear for fluorescent bodies
  - Study of sex chromatin or Barr bodies
  - Study of drumstick body in neutrophils.

KARYOTYPING

Q. Write a short note on karyotyping.
- Karyotype is a complete chromosome set of a cell.
- Karyotyping is a method of obtaining a karyotype of a cell.
- For chromosome analysis, the karyotype is prepared from the arrested cell division at metaphase. *(NEET)*

Protocol for Karyotyping

- *Sample collection:* Blood, chorionic villus sample, amniotic fluid, bone marrow or fetal blood is collected in heparinised vacutainer.
- *Culture setup:* Cells from collected sample are transferred to culture media (RPMI 1640 or TC 199) containing fetal bovine serum (nutrition), phytohemagglutinin (mitotic agent) and antibiotics (to prevent bacterial growth).
- *Incubation:* Cultured cells are incubated at 37°C for 72 hours to allow cells to undergo mitosis.
- *Colchicine block:* Small amount of colchicine is added to the culture to block cell division in mitosis.

- *Harvesting:* Cultured cells are washed with a hypotonic solution (to remove red blood cells and to swell all the cells). These cells are fixed with chilled fixative (a mixture of glacial acetic acid and methanol).
- *Slide preparation:* Slide is prepared from the drop of fixed cells.
- *Staining:* Slide is stained with Giemsa stain, G-banding **(Box 1)**, Q-banding, R banding, C-banding or nucleolus organiser region-staining. *(Viva)*
- *Karyotyping:* Stained slide is observed under a microscope and microphotographed, and chromosomes are arranged **(Fig. 1)**.

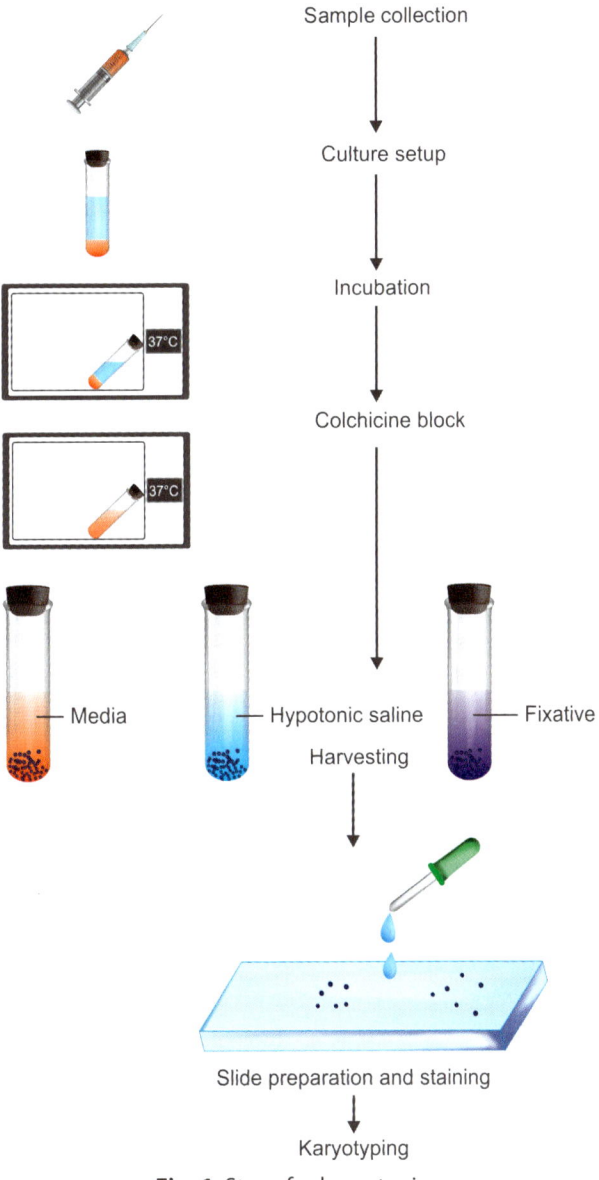

Fig. 1: Steps for karyotyping

Principles of Clinical Genetics

> **Box 1:** G-banding
>
> **G-banding**
> It is also known Giemsa banding. It is the **most common** employed procedure for karyotyping. (*NEET*)
>
> **Need of banding method**
> - To differentiate between chromosomes of the same group, for example, chromosomes 4 and 5.
> - To identify addition, deletion, inversion, and translocations.
>
> **Procedure**
> - Slide of chromosomes are treated with **trypsin** (to partially digest histone proteins) before Giemsa staining
>   ↓
> - Densely stained region of chromosome–**dark band**
> - Lightly stained region of chromosome–**light band**.

Preparation of Karyotype

- Photographed chromosomes are arranged in the following manner:
    - Shape of chromosome
    - Length of chromosome
    - Position of chromosome
    - Presence of satellite body.
- According to the **Denver classification**, chromosomes are arranged into seven groups A–G **(Table 1)**.
- While arranging karyotype, all chromosomes are arranged in pairs in their corresponding group, except X and Y that are arranged at the end (X is followed by Y).

Uses of Karyotyping

- Clinical diagnosis of cases with chromosomal aberrations.
- Clinical confirmation of cancer cases showing translocations, for example, detection of Philadelphia chromosome t(9;22)(q34;q11.2) in chronic myeloid leukemia.
- Prenatal diagnosis for chromosomal aberrations.
- Diagnosis of chromosomal defects in infertility cases or in recurrent abortion cases.

Note: For detection of single nucleotide change in DNA, polymerase chain reaction is used rather than cytogenetics.

Table 1: Denver classification of chromosomes

Group	Chromosome pair	Features
A	1, 2, 3	1 and 3– metacentric, 2–submetacentric
B	4, 5	Submetacentric
C	6–12, X (*NEET*)	Submetacentric
D	13–15	Acrocentric
E	16–18	16–metacentric, 17, 18–submetacentric
F	19–20	Metacentric
G	21–22, Y (*NEET*)	Acrocentric

Cytogenetics   13

Some Interesting Facts

Q. Enlist various methods for chromosome banding.

- *Q-banding:* In Q-banding, chromosomes are stained with fluorescent dye—quinacrine and it is useful for studying **heterochromatin** of chromosomes 1, 9 and 16.
- *R-banding:* In R-banding, slides are preheated before Giemsa staining to generate the reverse G-banding pattern.
- *C-banding:* In C-banding, DNA of the chromosome is denatured with acid, alkali or heat prior to Giemsa staining. It is useful for studying **constitutive heterochromatin**.

FLUORESCENT *IN SITU* HYBRIDIZATION

Q. Write short note on FISH.
- Fluorescent *in situ* hybridization (FISH) is a cytogenic technique that utilises the property of **DNA probe** to bind with its complementary target DNA sequence **(Box 2)**.
- DNA probes are tagged with fluorescent substances (may be with radioactive labels) *(NEET)*

Procedure

- *Metaphase FISH:* Cells collected from tissue → culture → colchicine block → harvesting → metaphase preparation → hybridisation with probe → observation under a fluorescent microscope.
- *Interphase FISH:* Cells collected from tissue → slide preparation → hybridisation with DNA probe → observation under a fluorescent microscope.

Applications of FISH

- To confirm the presence of specific chromosome or gene.
- To assess structural defects of chromosomes such as translocation.
- To identify small deletion or addition of chromosome.

Box 2: Feulgen reaction, SKY, HRB

Feulgen reaction
It is a staining method for cellular DNA that reacts with Schiff's reagent. It is specific for 2-deoxyribose sugar. *(NEET)*

Spectral karyotyping (SKY)
- It is a cytogenetic technique that involves simultaneous **painting of all human chromosome** with different fluorescent-colored DNA probe.
- SKY technique is useful for identification of **multiple translocations**.

High-resolution banding (HRB) technique
- In chromosome banding technique, total number of bands in all chromosomes is called as **band level**.
- Chromosomes in metaphase are condensed → hence can produce only 200–350 bands.
- For detailed analysis, longer chromosomes are obtained from **late prophase** or **early metaphase**.
- **Procedure:** Similar as karyotyping except treatment with ethidium bromide, actinomycin D or Hoechst with short exposure of colchicine.
- **Use:** To obtain 800–1400 band level for easy diagnosis of deletion, addition, translocation and many other chromosomal aberrations.

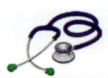

SOLUTION FOR CLINICAL CASE

As the referred case is suspected for chronic myeloid leukemia, the cytogenetic investigation is essential for the confirmation of the genetic basis of the disease. For this case, expected abnormality is the presence of translocation between chromosome 9 and 22 [t(9;22)(q34;q11.2)]. It can be detected by karyotyping with G-banding and confirmed by FISH. For karyotyping, 72 hours' culture setup is required; hence the clinician should not expect the immediate cytogenetic reporting.

4
CHAPTER

Chromosomal Aberration

CLINICAL CASE

A 16-year-old girl attended the gynecology OPD with complaints of primary amenorrhea and recurrent ear infections. On examination, the patient was short stature and had a webbed neck with a low hairline, low-set ears, shield chest and cubitus valgus. Investigations revealed increased follicle-stimulating hormone and low estradiol. On ultrasound, bilateral streak ovaries on both sides and hypoplastic uterus were seen. What investigation will you suggest for confirmation of your diagnosis?

INTRODUCTION

- Chromosomal aberration is a chromosomal abnormality that involves a missing, extra or irregular portion of a chromosome.
- Chromosomal aberrations are studied with the help of karyotype. Karyotype is a full set of chromosomes from a patient.
- Abnormal chromosome produces inherited abnormalities.

CLASSIFICATION OF CHROMOSOMAL ABERRATIONS

Chromosomal aberrations are classified into two groups:
1. Structural aberrations
2. Numerical aberrations.

STRUCTURAL CHROMOSOMAL ABERRATIONS

Q. Enlist the structural chromosomal aberrations.
- When the chromosomal structure is altered, it produces structural aberrations.
- Structural chromosomal aberrations may be stable or unstable.
- Structural chromosomal aberrations are classified as given in **Box 1**.

Deletion

Q. Write a short note on deletion.
- It is also called as deletion mutation.
- In deletion, part of a chromosome is lost during DNA replication.

Principles of Clinical Genetics

> **Box 1:** Classification of structural chromosomal aberrations
> - Deletion—terminal, interstitial, microdeletion
> - Translocation—Robertsonian, reciprocal
> - Insertion
> - Inversion—pericentric, paracentric
> - Isochromosome
> - Ring chromosome
> - Duplication.

- There are three types of deletion aberrations:
  1. *Terminal deletion:* A terminal segment of the chromosome is lost. For example, deletion of terminal part of the short arm of chromosome 5 → Cri-du-chat syndrome.
  2. *Interstitial or intercalary deletion:*
      - A segment of chromosome between two breaks is lost.
      - For example, a deletion of segment of long arm of chromosome 15 (15q11–13) → Prader-Willi syndrome. *(NEET)*
  3. *Microdeletion:*
      - It involves deletion of small segment of chromosome usually less than 5 million base pairs (5 mb).
      - For detection of microdeletion high-resolution banding karyotype or fluorescent in situ hybridisation (FISH) technique is required.
      - Examples of microdeletion are listed in **Table 1**.

Table 1: Examples of microdeletion *(NEET)*

Disease	Involved segment in deletion
Wolf-Hirschhorn syndrome	4p16.3
William syndrome	7q11.23
Langer-Giedion syndrome	8q24
Wilms tumor with aniridia	11q13
Prader-Willi syndrome **(Box 2)**	15q11–13
Angelman syndrome **(Box 2)**	15q11–13
Rubinstein-Taybi syndrome	16p13.3
Miller-Dieker syndrome	17p13.3
Smith-Magenis syndrome	17p11.2
DiGeorge syndrome	22q11.2
Few genes from these segments get deleted in the syndromes.	

> **Box 2:** Prader-Willi and Angelman syndromes
> - If the deletion of genes in segment 15q11–13 is inherited from father → Prader-Willi syndrome and from mother → Angelman syndrome. *(NEET)*

Chromosomal Aberration

Some Interesting Facts

Cri-du-chat Syndrome

Q. Write a short note on Cri-du-chat or Lejeune's syndrome.
- Also known as *chromosome 5p deletion* syndrome or 5p− syndrome or *Lejeune's* syndrome
- Cat-cry in French = Cri-du-chat.
- Characterized by cat-like cry of affected child.

Cause: Deletion of short arm of chromosome 5 (5p−)

Incidence: 1 in 50,000 live births

Clinical features
- Cat-like cry up to 2 years of age
- Epicanthus (skin fold of upper eyelid medial corner of eye)
- High-pitched voice
- Severe intellectual disability (due to *CTNND2* gene deletion)
- Microcephaly (small head)
- Micrognathia (small jaw)
- Low sets of ears
- Round face
- Developmental delay.

Genomic Imprinting

Q. Write short note on genomic imprinting.

Definition

- Genomic imprinting is a phenomenon that involves the expression of a gene depending upon its maternal or paternal origin **(Flow chart 1)**.
- Imparted gene is not expressed but similar gene (allele) carried from another parent is expressed. *Example:* Prader-Willi/Angelman syndrome.

Flow chart 1: Genomic imprinting

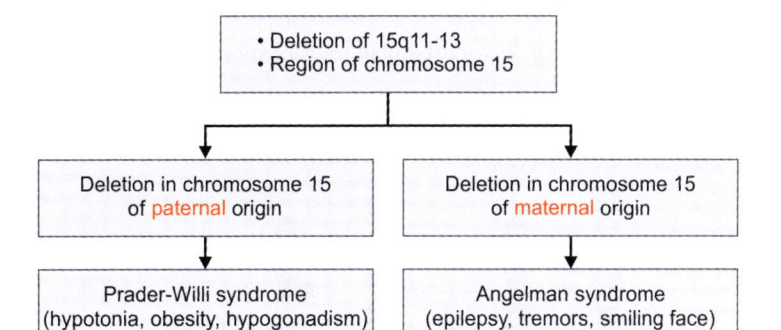

```
        • Deletion of 15q11-13
        • Region of chromosome 15
              │
      ┌───────┴───────┐
      ▼               ▼
Deletion in      Deletion in
chromosome 15    chromosome 15
of paternal      of maternal
origin           origin
      │               │
      ▼               ▼
Prader-Willi     Angelman
syndrome         syndrome
(hypotonia,      (epilepsy,
obesity,         tremors,
hypogonadism)    smiling face)
```

Mnemonic: P for paternal and Prader-Willi; Angelman and mother—she is our angel

Translocation

Q. Write a short note on translocation.
- Translocation is chromosomal abnormalities that occur when chromosomes break and the fragment rejoin to another nonhomologous chromosome.
- There are two types of translocations: Robertsonian and reciprocal translocations.

Robertsonian Translocation
- Robertsonian translocation involves two acrocentric chromosomes (**Fig. 2**). *(NEET)*
- It involves break at centromere and fusion of chromosome with the loss of acrocentric segments (short arms). *Examples:*
  - Robertsonian translocation between chromosome 13 and 14. (1 in 1000 people and most common Robertsonian translocation). *(NEET)*
  - An individual with Robertsonian translocations between two chromosome 21 may have a child with Down syndrome.

Reciprocal Translocation
- Exchange of material between two nonhomologous chromosomes (**Fig. 3**).
- Reciprocal translocation is a balanced translocation and there is no loss of chromosomal material.
- An individual with reciprocal translocation produce abnormal gametes and it results into spontaneous abortions.

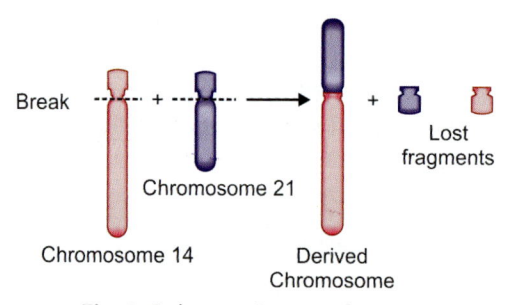

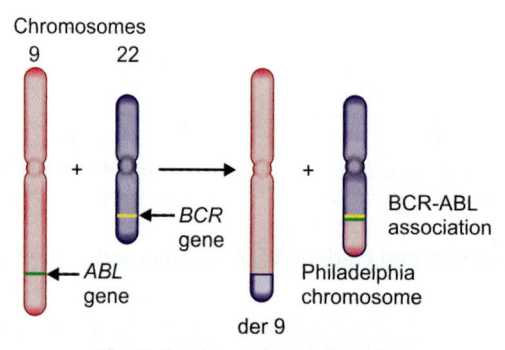

Fig. 2: Robertsonian translocation

Fig. 3: Reciprocal translocation

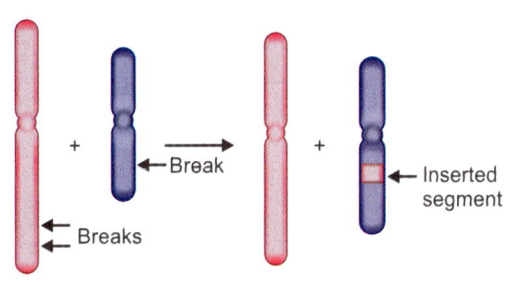

Fig. 4: Insertion mutation

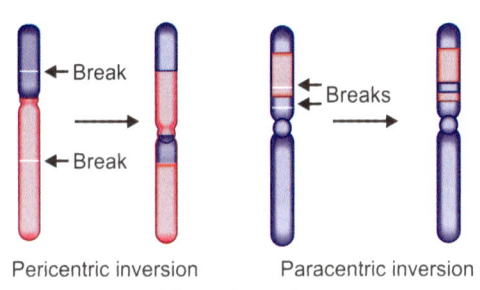

Pericentric inversion   Paracentric inversion
Fig. 5: Inversion

- *Example:* Philadelphia chromosome—it is produced by reciprocal translocation between chromosome 9 and 22.
- Philadelphia chromosome is associated with chronic myeloid leukemia (CML) (85% cases).

Insertion
Insertion is transposition of a segment of one chromosome to another chromosome **(Fig. 4)**.

Inversion
Q. Write a short note on inversion mutation.
- Inversion takes place when there are two breaks in a single chromosome and a segment turns 180° before the reunion.
- *Two types of inversion* **(Fig. 5)**:
  1. *Pericentric:* Breaks in the short and long arm of a chromosome, and the segment rotates around centromere → pericentric inversion.
  2. *Paracentric:* Breaks in the same arm of a chromosome (segment do not involve centromere) → segment rotates and fuse → paracentric inversion.

Isochromosome
- Transverse division of centromere → fusion of short arms with each other and fusion of long arms → isochromosomes **(Fig. 6)**.
- In derived chromosomes, arms are mirror images of each other.
- Isochromosome formation in the acrocentric or submetacentric chromosome may lead to the loss of genetic material.
- For example, some cases of Turner syndrome have i(Xq)—isochromosome of the long arm of chromosome X with the loss of short arms.

Principles of Clinical Genetics

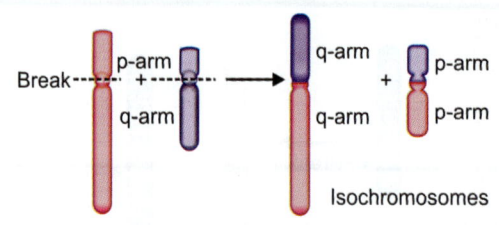

Fig. 6: Isochromosomes

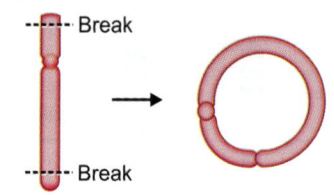

Fig. 7: Ring chromosome

Ring Chromosome
- Two breaks in a chromosome at its end → fusion of sticky ends and loss of material distal to breaks → ring chromosome **(Fig. 7)**.
- For example, ring X chromosome [r(X)] may produce Turner syndrome.

Duplication
Chromosomal duplication involves the presence of a portion of a chromosome more than once in a gamete or more than twice in a zygote.

FACTORS CAUSING CHROMOSOMAL ABERRATIONS
The factors playing important role in chromosomal aberrations are listed in **Box 3**.

Box 3: Factors causing chromosomal aberrations
- Advanced maternal age
- Radiation
- Parental chromosomal abnormalities
- Autoimmune disorders
- Nutritional deficiencies (folic acid)
- Alcohol, tobacco, drug exposure
- Infection (viruses and bacteria).

NUMERICAL CHROMOSOMAL ABNORMALITIES
- Normally, human diploid cells contain 46 chromosomes (46,XY in male and 46,XX in female)
- Any deviations in total number of chromosome are known as numerical chromosomal aberrations.

Some Definitions

- *Ploidy:* Number of sets of chromosomes in a cell.
- *Haploid:* One set of chromosomes. For example, ovum (23,X) or sperm (23,Y)
- *Diploid:* Cells with two sets of chromosomes. For example, somatic human cells (46,XY or 46,XX).
- *Polyploid:* Cells with more than two sets of chromosomes, for example, triploid (69 chromosomes), tetraploid (92 chromosomes).
- *Homoploid:* Cells with same number of chromosomes.
- *Aneuploidy:* Cells with abnormal number of chromosomes that is not multiple of haploid set. For example, 2n + 1, that is, 47 chromosomes and 2n – 1, that is, 45 chromosomes. *(Viva)*
- *Nondisjunction:* It is the failure of separation of homologous chromosomes or sister chromatids during cell division **(Fig. 8 and Table 2)**.

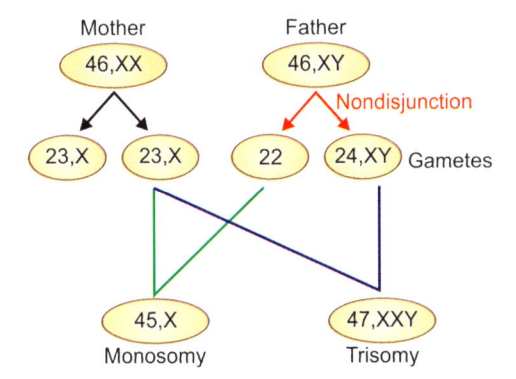

Fig. 8: Mechanism of nondisjunction

Table 2: Examples of nondisjunction

Syndrome	Alternate name	Karyotype
Monosomy	Turner syndrome	45,X
Trisomy:		
Trisomy 21	Down syndrome	47,XY,+21
Trisomy 18	Edward syndrome	47, XY,+18
Trisomy 13	Patau syndrome	47, XY,+13
–	Klinefelter syndrome	47, XXY
Trisomy X	Triple X syndrome	47, XXX
Trisomy 8	Warkany syndrome	47, XX,+8
Tetrasomy:		
Tetrasomy 22	Cat eye syndrome	48, XX,+22+22
Tetrasomy 12p	Pallister-Killian syndrome	48, XX, +12p, +12p
Tetrasomy X		48, XXXX

Down Syndrome

Q. Write a short note on Down syndrome.
- It is also known as trisomy 21 or mongolism.
- First patient was identified by John Langdon Down (1866).
- *Karyotype:* 47,XY; +21 or 47,XX; +21
- *Dermatoglyphics:* Single transverse palmar crease—known as simian crease.
- *Incidence:* 1 in 800 newborns.

Clinical features **(Fig. 9)** *(Viva)*
- Mental retardation, decreased intelligent quotient.
- *Mongolian features:*
  - Round face
  - Almond-shaped eyes
  - Epicanthal folds
  - Slanting palpebral fissure
  - Low sets of ears
  - Low bridge of nose
  - Small chin (micrognathia)
  - Macroglossia

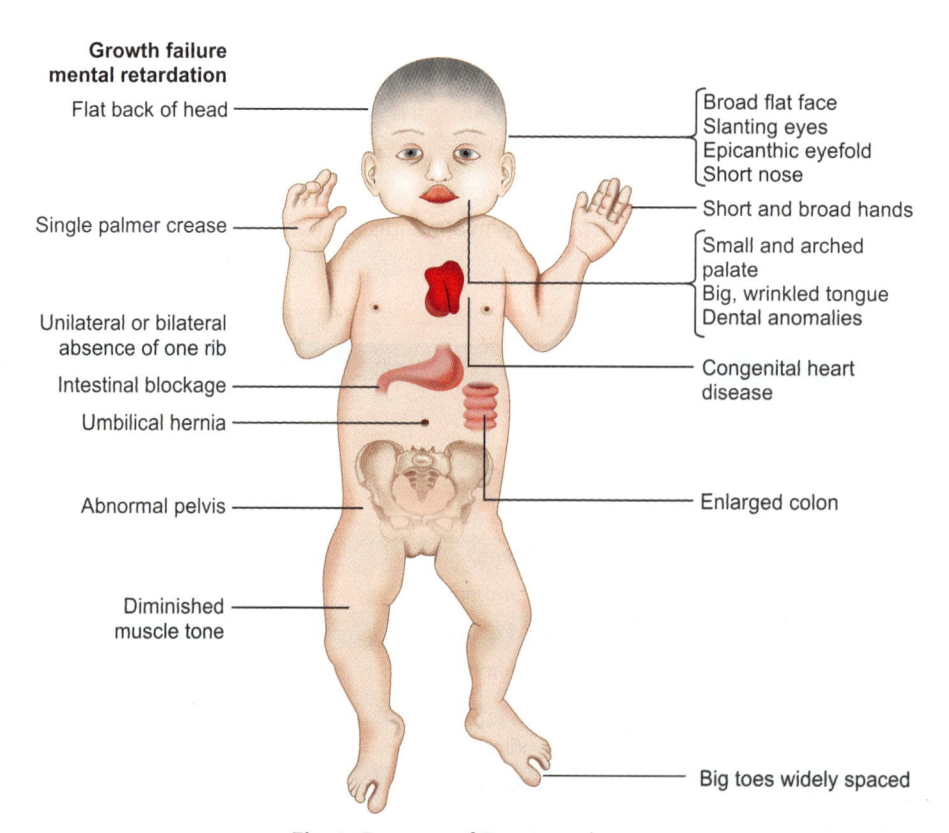

Fig. 9: Features of Down syndrome

Chromosomal Aberration

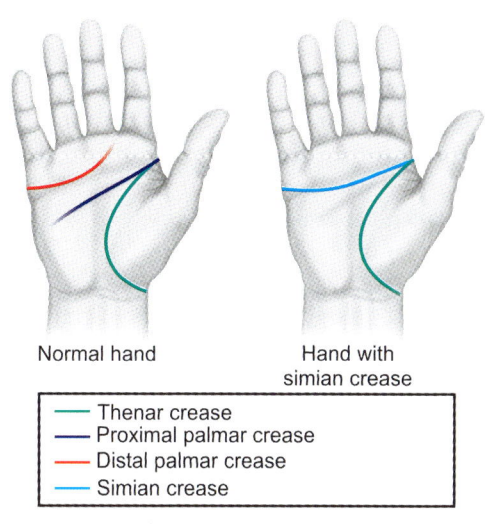

Fig. 10: Simian crease

- Hypotonia
- Simian crease **(Fig. 10)**
- Larger space between big toe and second toe
- Incurving of the fifth finger (clinodactyly).

Causes
- Increased maternal age (specifically after 35 years of age) *(NEET)*
- Chromosomal nondisjunction

Prenatal testing
1. In maternal serum at 16–18 weeks of pregnancy, if the fetus has Down syndrome, it is marked by: *(NEET)*
   a. Reduced α-fetoprotein (less than 25%).
   b. Reduced unconjugated estradiol (less than 25%).
   c. Increased human chorionic gonadotropin level (more than 200%).
2. Chorionic villus sampling or amniocentesis for karyotyping.

Some Interesting Facts
- In Down syndrome, *DYRK1*, a kinase gene located on the long arm of chromosome 21 produces mental retardation. *(NEET)*
- Life expectancy in Down syndrome is increased to 60 years due to health sector developments.

Turner Syndrome

Q. Write a short note on Turner syndrome.
- It is also known as ovarian dysgenesis or X monosomy or 45X0 syndrome or Ullrich-Turner syndrome.
- Described first by Henry Turner (1938).

Karyotype: 45,X

Phenotype:

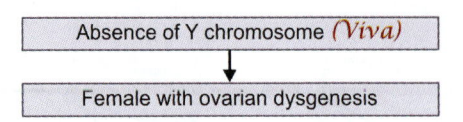

Absence of Y chromosome *(Viva)*
↓
Female with ovarian dysgenesis

Incidence: 1 in every 2,500–3,000 newborn girls.

Clinical features **(Fig. 11)** *(Viva)*
- Female with gonadal dysgenesis → streak-like gonads → no ovarian follicles → no sexual hormones → primary amenorrhea (no menstruation) → infertility
- Short stature
- Webbing of neck (skin folds in neck)
- Cubitus valgus (decreased carrying angle at elbow)
- Broad chest, widely placed nipples, underdeveloped breast
- Lymphedema over limbs (swollen)
- A low hairline
- No secondary sexual character development.

Barr body: absent. *(NEET)*

Other associated features
- High-arched palate
- Visual abnormalities (strabismus)
- Coarctation of aorta, ventricular septal defects

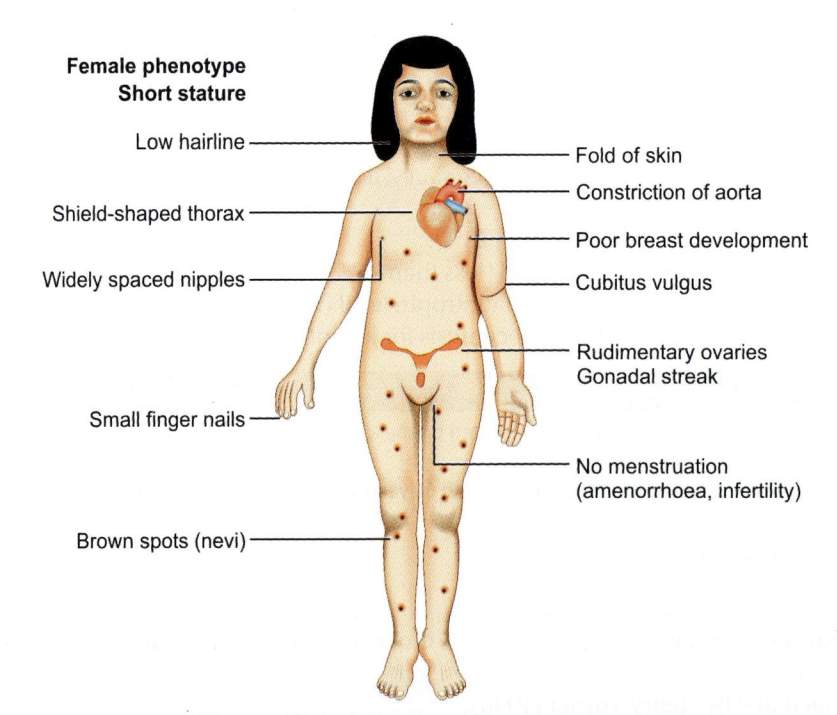

Fig. 11: Clinical features of Turner syndrome

- Small defects
- Horseshoe kidney
- Scoliosis
- Lack of estrogen → Osteoporosis.

Cause: Chromosomal nondisjunction during gametogenesis.

Investigations
- *Buccal smear:* Absence of Barr body.
- *Blood/serum:* Low levels of estrogen, progesterone.
- *Ultrasonography:* Gonadal absence
- *Karyotyping:* 45,X or mosaicism (46,XX/45,X).
- *Prenatal diagnosis:* Amniocentesis and chorionic villus sampling.

Treatment
Growth hormone and estrogen replacement therapy.

Some Interesting Facts

Mutation of *SHOX* gene located at distal tip of chromosomes Xp and Yp → short stature. *(NEET)*

Klinefelter Syndrome

Q. Write a short note on Klinefelter syndrome.
- It is also known as 47,XXY syndrome.
- First described by Harry Klinefelter (1942).

Karyotype: 47,XXY

Phenotype: Male

Incidence: 1 in 1,000 newborn boys.

Clinical features **(Fig. 12)** *(Viva)*
- Phenotypically male
- Small testis (hypogonadism) → decreased secretion of testosterone → delayed puberty → underdeveloped secondary sexual character, infertility and azoospermia.
- Gynecomastia (enlarged breast)
- Sparse body hairs
- Wide lips
- Taurodontic teeth (enlarged body of the tooth with small root)
- Long arms and legs
- Small penis
- Speech and language problems
- Osteoporosis
- Shyness and difficulty in expressing feeling

Cause: Nondisjunction during gametogenesis.

Investigations
- *Testicular biopsy:* Hyalinisation of seminiferous tubules; no spermatogenesis.
- *Buccal smear:* Barr body present.
- *Serum:* Low level of testosterone and increased luteinising hormone (LH) and follicle-stimulating hormone (FSH) levels.

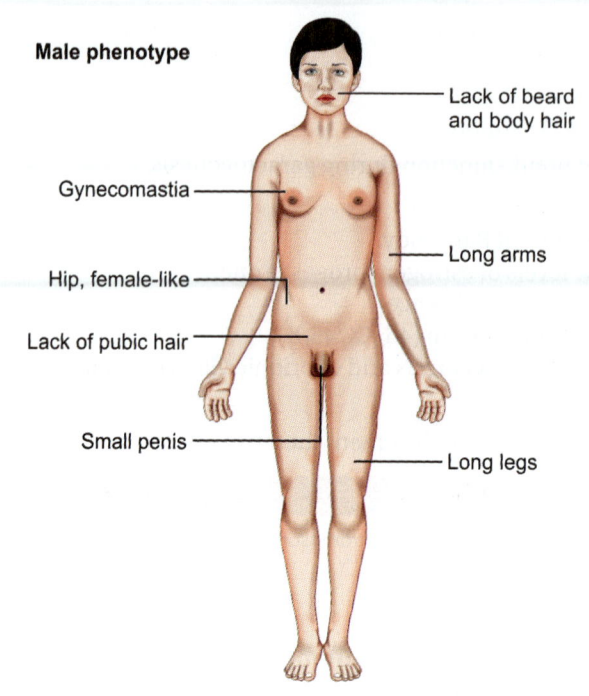

Fig. 12: Clinical features of Klinefelter syndrome

- *Karyotype:* 47,XXY.
- *Treatment:* Testosterone therapy.
- For infertility cases, few sperms collected from testicular biopsy generated successful pregnancies using *in vitro* fertilization (IVF).

Patau Syndrome

Q. Write a short note on Patau syndrome.
- It is also known as trisomy 13 or trisomy D.
- Mostly 80% of children with Patau syndrome die within first year of life.
- First described by Klaus Patau (1960).

Karyotype: 47,XY,+13 or 47,XX,+13

Cause: Nondisjunction during gametogenesis.

Clinical features
- The cleft clip, cleft palate
- Polydactyly (extra finger or toe)
- Microcephaly (small head)
- Sensory nystagmus
- Mental retardation
- Overlapping fingers
- Simian crease
- Rocker bottom feet (congenital vertical talus → prominent heel due to projected calcaneus)

- *Peters anomaly:* It is anterior segment mesenchymal dysgenesis causing abnormalities of cornea. *(NEET)*
- Microphthalmia (small eye)
- Holoprosencephaly (failure of forebrain development)
- Cyclopia (Single eye due to failure of forebrain separation).

Edward Syndrome

Q. Write a short note on Edward syndrome.
- It is also known as trisomy 18 or trisomy E.
- First described by John Edward (1960).
- It is the second most common autosomal trisomy (the first one is Down syndrome). *(NEET)*

Karyotype: 47,XY,+18 or 47,XX,+18.

Incidence: 1 in 6,000 live births. Fetuses mostly abort (95% cases), die in first few months after birth.

Cause: Nondisjunction during gametogenesis.

Clinical features
- Small head (microcephaly), low sets of ears
- Small jaw and mouth.
- Clenched fists with overlapping fingers.
- Rocker bottom foot.
- Associated respiratory, cardiovascular and genitourinary system malformations.

> **Some Interesting Facts**
>
> **XYY Syndrome**
> - Male phenotypic individuals with one extra Y chromosome (47,XYY).
> - They show behavioral disturbances, specifically may be antisocial behavior.
> - *Karyotyping:* 47,XYY. Quinacrine staining—double fluorescent spot in somatic cells.

Hermaphroditism

Q. Write a short note on hermaphroditism.
- Difficult differentiation of external genitalia as male or female.
- *True hermaphroditism (ovotesticular disorder):* Ambiguous external genitals and presence of both testis and ovaries.
- *Pseudohermaphroditism:* An individual with gonads of one sex and secondary sexual characters of another sex. For example, female pseudohermaphrodite—female with ovary (autosomal recessive disorder), adrenal hyperplasia → increased secretion of androgens → male secondary sexual characters.

Chimeras
- An individual composed of cells from different zygotes. *(Viva)*
- It may occur due to the fusion of multiple fertilized zygotes.
- In human, dispermic chimera and blood group chimera have been reported.

Mixed twins
- Mixed twins are dizygotic (fraternal) twins born to multiracial families. These twins differ in skin color and other racial features from each other.

Box 4: Fragile X syndrome

Fragile X syndrome

Q. Write a short note on fragile X syndrome.
- Caused by the fragile site on the long arm of chromosome X → breaks near the tip of long arm in laboratory testing.
- More common in males.
- Clinical features
  - Mental retardation
  - Autism
  - Typical facial features (protruding ears, long face)
  - High-arched palate
  - Flat feet
  - Hyperextensible fingers and thumb.
- *Cytogenetics:* Mutation in *FMR1* (forgive X mental retardation gene) in X chromosome.

SOLUTION FOR CLINICAL CASE

This is a typical presentation of Turner syndrome.
- For confirmation of the diagnosis, following investigations may be advised:
- Karyotyping (expected result: 45,X)
- Buccal smear (No Barr body)
- *Echocardiography:* To rule out associated heart diseases.

On diagnosis, the patient should be managed symptomatically and should be referred for genetic counselling for advice on marriage, reproduction and lifestyle changes.

5
CHAPTER

Structure of DNA and RNA

CLINICAL CASE

A 20-year-old male with the history of unilateral ptosis for the past 6 months and bilateral ptosis for the past 2 weeks, progressive exercise intolerance and bradycardia. Similar clinical history was found in his mother who is no more. The clinician called the genetic counsellor for help. What should be the response given by the genetic counsellor?

INTRODUCTION

Nucleotides
- DNA and RNA consist of nucleotides that form the building blocks.
- Nucleotide has three components:
  1. *Nitrogenous base:* May be purine (adenine and guanine) or pyrimidine (cytosine, thymine and uracil).
     Note: Pyrimidines are also found in thiamin. *(NEET)*
  2. *Pentose sugar:* Ribose in RNA and deoxyribose in DNA.
  3. *Phosphate molecules:* May be one to three phosphate molecules.

> **Box 1:** Facts about DNA
>
> *Facts about DNA:*
> - Xanthine and hypoxanthine are also purine bases but are not present in nucleic acids.
> - Nucleoside = Nitrogen base + Pentose sugar.
> - Nucleotide = Nitrogen base + Pentose sugar + phosphate molecule.
> - One kilobase (1 kb) length of nucleic acid consists of 1000 base pairs in double-stranded and 1,000 bases in single-stranded nucleic acid. *(NEET)*
> - New DNA synthesis occurs in interphase of cell division. *(NEET)*

DEOXYRIBONUCLEIC ACID
- Deoxyribonucleic acid (DNA) is responsible for heredity and expression of characters.
- DNA is a polymer of deoxyribonucleotides.
- DNA consists of adenine, guanine, cytosine, thymine, deoxyribose sugar and phosphate molecules.

Watson and Crick Model

- DNA is a spiral twisting of two polyribonucleotide chains in the form of right-handed double helix structure **(Fig. 1)**.
- Both chains run in an antiparallel direction (one 3′ to 5′ and another 5′ to 3′). Base sequence in DNA molecule is always written from 5′ to 3′. *(NEET)*
- Both the chains are held together by hydrogen bonds between nitrogen bases.
- Adenine (A) pairs with thymine (T) by two hydrogen bonds and cytosine (C) pairs with guanine (G) by three hydrogen bonds. Three hydrogen bonds between G and C provide *thermal stability* to DNA. *(NEET)*

Box 2: Chargaff's rule

Chargaff's rule
- *Chargaff's rule:* For double-stranded DNA: A + G (purines) = C + T (pyrimidines). *(NEET)*
- If the DNA contains A + G ≠ C + T, then the DNA is single-stranded. *(NEET)*

Keys: A = adenine, T = thymine, C = cytosine, G = guanine

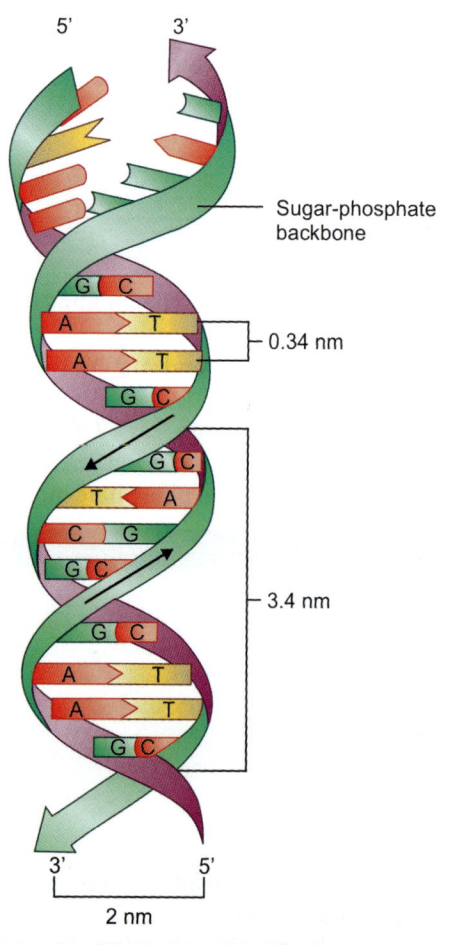

Fig. 1: Structure of DNA

Keys: A = adenine, T = thymine, C = cytosine, G = guanine

Box 3: DNA specification

DNA specification
- The length of each of helix: 3.4 nm.
- The width of helix: 2 nm.
- Each helix = 10 base pairs.
- The distance between adjacent base pairs: 3.4 nm.

RIBONUCLEIC ACID

- Ribonucleic acid (RNA) is a polymer of ribonucleotides that consists of the ribose sugar, adenine, guanine, cytosine, *uracil* and phosphate molecules.
- *There are three types of RNA:* Messenger RNA (mRNA), ribosomal RNA (rRNA) and transfer RNA (tRNA).

Transfer RNA (tRNA)

- Constitutes 20% of total RNA.
- Carrier of amino acids to ribosomes in protein synthesis.
- Each tRNA is specific for amino acid (some amino acids can be carried by more than one tRNA)
- An average number of nucleotides in tRNA varies between *74 to 95 nucleotides.* (*NEET*)
- The tRNA contains unusual bases such as pseudouracil, thymidine. (*NEET*)
- The tRNA has five arms as follows **(Fig. 2)**:
  1. *Acceptor (CCA) arm:* It has 3' end with base sequence cytosine-cytosine-adenine. It binds with amino acids.
  2. *TΨC arm:* It binds tRNA to the ribosome.
  3. *Variable arm:* Useful for species' identification.
  4. *Anticodon arm:* It has *anticodons* and it binds with specific codon of mRNA during translation. (*NEET*)
  5. *D arm:* It acts like recognition site for aminoacyl-tRNA synthetase enzyme that adds specific amino acids to tRNA.

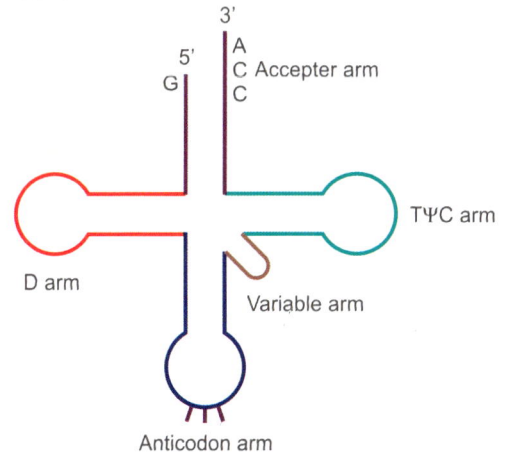

Fig. 2: Transfer RNA

Keys: Ψ = pseudouracil, A = adenine, C = cytosine, G = guanine

Ribosomal RNA

- Constitutes 60%–70% of total RNA.
- rRNA combines with proteins to form ribosomes.
- rRNA is produced in the *nucleolus*. *(NEET)*
- Each ribosome has two subunits: larger subunit (60S) and smaller subunit (40S).
- *Function:* It helps in protein synthesis by providing a site for interaction between mRNA and tRNA.
- The rRNA have peptidyl transferase activity. *(NEET)*

Messenger RNA

- mRNA is synthesized in the nucleus from DNA by transcription. mRNA is a complimentary copy of the single-stranded DNA. *(NEET)*
- Synthesized mRNA moves to the cytoplasm and at the ribosome, it gets translated to form protein.

Structure *(Fig. 3)*

- mRNA has 7-methyl GTP cap at 5′ end (for protection from 5′ exonuclease). *(NEET)*
- mRNA has poly-A tail of 22–250 AMP at 3′ end (for stability and protection from 3′ exonuclease action). *(NEET)*
- mRNA has codons (sequence of *three bases*) for identification of specific anticodons of tRNA. *(NEET)*

Some Interesting Facts *(NEET)*

- A, U, G and C bases in mRNA form *64 triplets (codons).*
- *61 triplets* are codons as they recognise tRNA anticodon.
- Three triplets (*UAG, UGA, UAA*) are *nonsense codons* or chain *termination* codons as they cannot recognise any tRNA.
- The three stop codons have been given names: UAG is amber, UGA is opal (sometimes also called umber), and UAA is ochre.
- AUG codon is chain initiation codon.
- Methionine and tryptophan have only one codon for each of them.
- Reverse transcriptase converts single-stranded DNA into double-stranded RNA.
- Degeneracy of codon indicates more than one codon for single amino acid. Genetic code cannot be universal or overlapping.
- RNAs are less stable than DNA due to the presence of two hydroxyl groups. Hence, DNA is selected for genetic information.
- If there is an insertion in the region of the intron of the DNA, it will not be expressed and the resultant protein will be normal.
- DNA synthesis occurs only in S phase, whereas RNA and protein synthesis occurs in all phases of cell cycle.

Fig. 3: Messenger RNA
Abbreviations: GTP, Guanosine triphosphate; A, Adenosine

DIFFERENCES BETWEEN DNA AND RNA (TABLE 1)

Q. **Write the differences between DNA and RNA.**

Table 1: Differences between DNA and RNA

DNA	RNA
• Double stranded	• Single stranded
• Present in nucleus or mitochondrion	• Present in the cytoplasm
• Sugar: Deoxyribose	• Sugar: Ribose
• Base pairs: A, T, G, C	• Base pairs: A, U, G, C
• Chargaff's rule follows (A = T, G = C)	• Does not follow Chargaff's rule
• Self-replicating	• Synthesised from DNA
• Alkali resistant	• Destroyed by alkali

SATELLITE DNA

- It is noncoding repeat DNA sequence.
- In the human satellite, DNA is present in heterochromatin region of chromosome 1, 9, 16, long arm of chromosome Y, and satellite bodies of chromosome 13, 14, 15, 21 and 22.
- It is responsible for coding ribosomal and transfer RNAs.

MITOCHONDRIAL DNA

Q. **Short note on mitochondrial DNA (mtDNA).**

- mtDNA is located in the mitochondrion and it constitutes approximately 1% of total cellular DNA. *(NEET)*
- All humans receive mtDNA from the mother as the only ovum provides cytoplasm and mitochondrion to the zygote (embryo). *(NEET)*
- *Evolution:* mtDNA entered in eukaryotic cells through bacterial infection (endosymbiotic theory of evolution).
- mtDNA is circular double-stranded DNA. *(NEET)*
- mtDNA codes for adenosine triphosphate synthase, cytochrome C oxidase, cytochrome B, NADH dehydrogenase.
- DNA polymerase γ is required for replication of mtDNA. *(NEET)*
- Mutations:

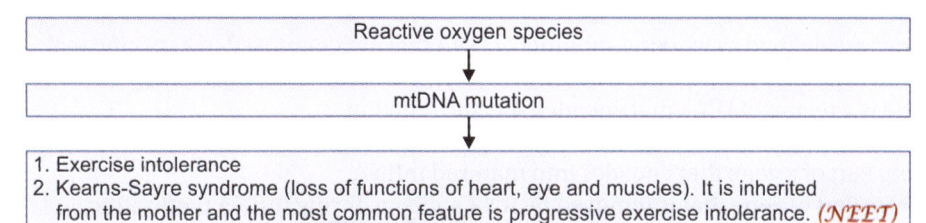

```
Reactive oxygen species
         ↓
    mtDNA mutation
         ↓
1. Exercise intolerance
2. Kearns-Sayre syndrome (loss of functions of heart, eye and muscles). It is inherited
   from the mother and the most common feature is progressive exercise intolerance. (NEET)
```

TRANSCRIPTION AND TRANSLATION

Transcription
Process of mRNA synthesis from DNA.

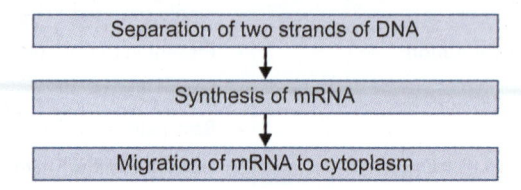

Process

```
Separation of two strands of DNA
           ↓
     Synthesis of mRNA
           ↓
Migration of mRNA to cytoplasm
```

Translation
The process of protein synthesis from mRNA.

Process

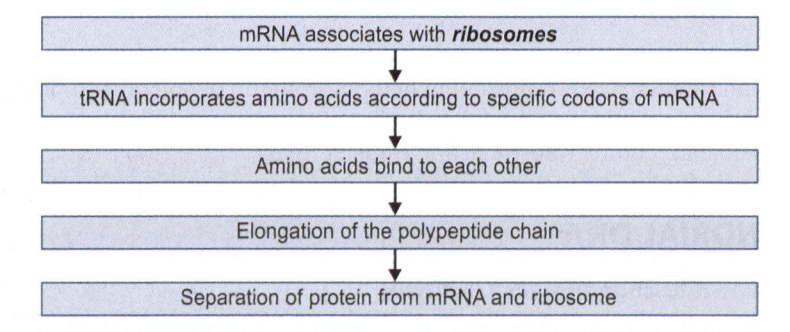

```
        mRNA associates with ribosomes
                    ↓
tRNA incorporates amino acids according to specific codons of mRNA
                    ↓
           Amino acids bind to each other
                    ↓
         Elongation of the polypeptide chain
                    ↓
     Separation of protein from mRNA and ribosome
```

> **Box 4:** Facts about translation
>
> *Facts about translation*
> - Ribozymes are RNA fragment with catalytic activity. *(NEET)*
> - Kozak sequence occurs in eukaryotic DNA and plays a role in translation. *(NEET)*
> - Aminoacyl-tRNA synthetase is fidelity enzyme in protein synthesis. It is required for the attachment of specific amino acids to the corresponding tRNA. *(NEET)*

GENE

Q. Short note on the gene.
- Genes are defined as working subunits of DNA that are composed of a specific sequence of codons and may get expressed in the phenotypic form (protein).
- Gene is a region of DNA that encodes specific function.
- Gene = exons + introns **(Fig. 4)**
- *Exon:* Part of a gene that encodes into matured mRNA.
- *Intron:* Part of gene that is removed by RNA splicing during mRNA maturation.
- Human chromosomes have 30,000 genes.
- Genes are either structural genes (produces proteins) or regulatory genes (promote or inhibit activity of other genes).

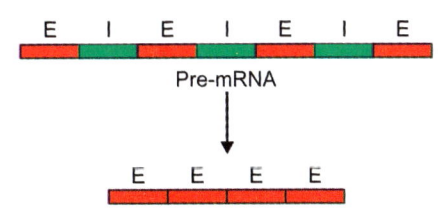

Fig. 4: Introns and exons

Genotype

Genes made up of different DNA sequences (total genes of a cell) are called genotype.

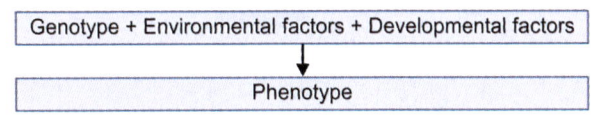

Genotype + Environmental factors + Developmental factors → Phenotype

Flanking Region

- It is a region of DNA that contains promoter gene and a termination codon.
- It is essential for the beginning of mRNA synthesis, but it does not get transcribed into mRNA.
- It is also essential for the termination of protein synthesis.
- At 5' end, flanking possess promoter zone with TATA box and CAT box.
- TATA box is called as Goldberg-Hogness box. It is the region that helps in transcription. It has 5'TATA........TAA 3' DNA sequence.
- CAT box is also called as CCAAT box. It also helps in transcription.
- At 3' end, the flanking region possesses termination codon (TAA).

Transcription

- It is the process of mRNA synthesis.
- Actinomycin D inhibits transcription. *(NEET)*
- Steps **(Fig. 5)**:

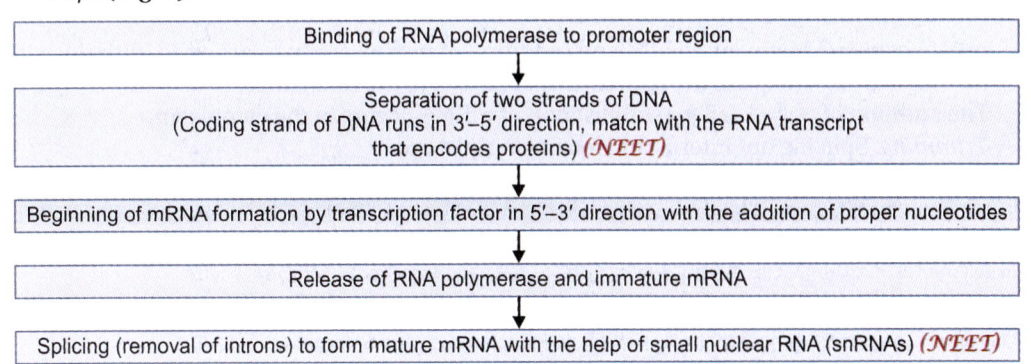

Binding of RNA polymerase to promoter region
↓
Separation of two strands of DNA
(Coding strand of DNA runs in 3'–5' direction, match with the RNA transcript that encodes proteins) *(NEET)*
↓
Beginning of mRNA formation by transcription factor in 5'–3' direction with the addition of proper nucleotides
↓
Release of RNA polymerase and immature mRNA
↓
Splicing (removal of introns) to form mature mRNA with the help of small nuclear RNA (snRNAs) *(NEET)*

- *Poly-A tail:* It is present at 3' end of mRNA degradation and prevents mRNA degradation.
- *Template strand:* It is a DNA strand (3' to 5') that forms mRNA.
- *Transcription factor:* Protein that attracts RNA polymerase to the promoter region.

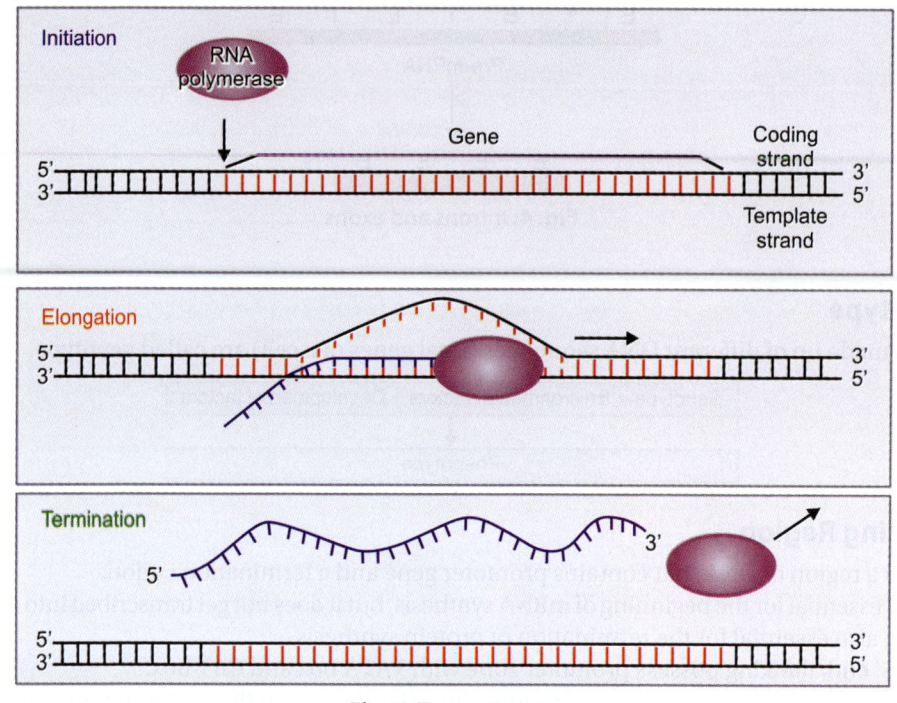

Fig. 5: Transcription

- The intron is the segment of the gene that is not represented in the matured mRNA. *(NEET)*
- Cistron is the smallest fundamental unit coding for the DNA synthesis. *(NEET)*
- Gene = intron + exon; and Cistron = all exons of a single gene.
- Replication and transcription are the similar processes as these involve the formation of the phosphodiester bonds in elongation of the chain in 5′–3′ direction. *(NEET)*

Processing of mRNA (Post-transcriptional Modifications) *(NEET)*
- *mRNA capping:* Methylation of 5′ end (addition of methyl group)—act as an initiation site for translation. It helps in the attachment of mRNA to the ribosome.
- The addition of poly-A tail for stabilisation of mRNA occurs in the cytoplasm.
- *Trimming:* Splicing out interns and leaving only exons.

Box 5: Initiation factors and alternate splicing

Initiation factors
In eukaryotes, initiation factor is regulated by GTP to GDP transformation. *(NEET)*

Alternate splicing
One gene can synthesise more than one protein due to alternate splicing **(Fig. 6)**.

RNA Polymerase *(NEET)*
- It is DNA-dependent RNA polymerase.
- It produces primary transcript RNA from DNA.

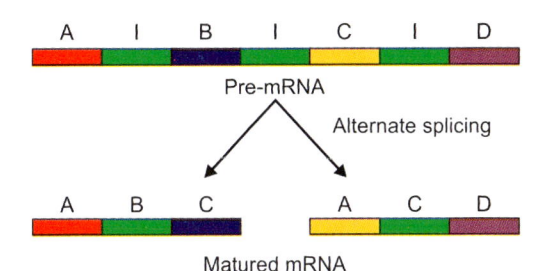

Fig. 6: Alternate splicing

- It has alpha (for initiation of transcription), beta (for initiation of elongation), beta 1 (for non-specific DNA binding) and sigma (for promoter binding) subunits.

Translation
- It is the process of protein synthesis from mRNA.
- *Process:*
  - *Initiation:*
    - The 5' end of mRNA and initiator tRNA attaches to the small subunit of the ribosome.
    - The start codon is AUG (anticodon on tRNA is UAC and carry methionine). *(NEET)*
    - Large ribosomal subunit attaches to small subunit.
  - *Elongation:*
    - Small subunit of one codon further moves on mRNA and new tRNA binds with mRNA.
    - Peptidyl transferase forms a peptide bond between new and previous amino acid and soon large subunit also moves towards small one.
  - *Termination:*
    - The process of elongation continues till small unit reaches to the termination codon.
    - Polypeptide chain gets detached from ribosome and forms folding to develop tertiary protein structure.
- *Post-translational modifications (NEET):* It includes:
  - Trimming
  - Glycosylation
  - Covalent alteration
  - Phosphorylation
  - Hydroxylation.

Factors Regulating of Gene Expression
- *Operator gene:* Controls the expression of structural genes that lies adjacent to operator genes on some chromosome.
  Operon = Operator gene + Structural genes
- *Regulator genes:* Controls operator genes through repressor genes
- *Transcription factors:* Controls binding of RNA polymerase at promoter genes. Transcription factor may be enhancer (increases transcription) or silencer (decreases transcription).
- For lac operon model, lactose, allolactose, isopropyl thiogalactoside, low glucose and CAP-cAMP are positive regulators or inducers. *LacI* gene and high glucose concentration are negative regulators or repressors. *(NEET)*

MUTATION

Q. Write a short note on mutation.

Definition
- Mutation is a heritable change in the gene.
- The rate of mutation in the human genome is one mutation per 10^6 bps. *(NEET)*

Mutagens: The agent that induces mutation is called mutagen.

Classification
Mutations are classified in different ways **(Table 2)**.

Causes of Mutation
- Spontaneous
- Radiation
- Ultraviolet rays
- X-rays
- α-rays, β-rays
- Temperature
- Chemical exposure
- Caffeine
- Formaldehyde.

Some Interesting Facts
- A mutation involving single gene = point mutation.
- A mutation involving chromosome = chromosomal mutation.
- Mutation can occur spontaneously (natural) or on exposure to mutagens (induced).
- A mutation may involve loss (deletion), addition (insertion) or multiplication (duplication) of segment of gene.
- If any mutation changes reading frame (codons) of the gene, it is called frameshift mutation.
- If a point mutation results into the substitution of one amino acid with another, it is called missense mutation.
- If point mutation results in premature stop codon or nonsense codon, such mutation is called nonsense mutation.

Box 6: Somatic and germ cell mutations
Somatic and germ cell mutations
- Somatic mutations involve somatic cells and are not transmitted to the offspring.
- Germ cell mutations involve gametes and are transmitted to the offspring.

Box 7: DNA repair
DNA repair
- Human cell repair spontaneous DNA damages using the following enzymes:
  - *DNA ligase:* Repairs a nick in DNA.
  - *AP endonuclease:* Replaces the lost base.
  - *Endonuclease, exonuclease, DNA polymerase, DNA ligase, photolyase:* Repair large DNA damage. *(NEET)*

Table 2: Classification of mutations

Basis	Mutation
Involved genes	Point mutation Chromosomal mutation
Origin	Natural Induced
Types of change	Structural Insertion Deletion Duplication Frameshift Missense Nonsense
Cell type	Somatic cell mutation Germ cell mutation

Some Interesting Facts

Jumping Genes/Transposons (NEET)
- Jumping genes can jump to and fro within a single chromosome or on adjacent chromosome.
- For example, *Alu* gene in human chromosomes.

Sister Chromatid Exchange
- Sister chromatid exchange (SCE) is an exchange of genetic material between sister chromatids
- 4–5 SCEs per chromosome pair per mitosis is normal.
- Bloom syndrome (Bloom-Torre-Machacek syndrome) is an autosomal recessive disorder with defective DNA helicase protein → genomic instability → increased SCE. *(NEET)*

Pseudogenes
Pseudogenes are nearly similar to some of the functional gene but pseudogenes are not able to produce proteins due to the abnormal regulatory region.

Processed or Retrotransposed Pseudogenes
A portion of mRNA is spontaneously reverse transcribed back into DNA and inserted into main chromosomal DNA. This insert is called processed pseudogenes. It does not contain introns.

Chain Termination Mutation
Mutation sometime converts codon into termination codon. Such mutation results into the premature termination of proteins.

Splice Mutation
A mutation affecting splicing of exons from introns is called splice mutation.

GENE MAPPING

Q. Write a short note on gene mapping.

Gene mapping is the creation of a genetic map, taking DNA fragments and assigning them to different chromosomes.

Definition
It is a method used to identify the locus of the gene and distances between genes.

Gene mapping is divided into two groups:
1. *Genetic mapping:* Uses the technique to construct the positions of genes, for example, pedigree charting.
2. *Physical mapping:* Uses the molecular biology techniques to examine DNA molecules directly.

For gene mapping, genetic markers are required.

Genetic Marker
- It is a gene or DNA sequence with a known location on a chromosome.
- It is useful to identify the presence of a gene and mutation in the gene to establish a relationship between the disease and genetic mutation.

Types of Genetic Markers
- Restriction fragment length polymorphism (RFLP)
- Amplified fragment length polymorphism (AFLP)
- Single nucleotide polymorphism (SNP) and so on.

Physical Mapping
It utilizes genetic markers for one of the following methods:
1. *Restriction mapping:* It locates the position of recognisable sequences for restriction endonucleases on a DNA molecule.
2. *Fluorescent in situ hybridisation (FISH):* It locates marker on a chromosome by direct hybridisation of the probe to DNA.
3. *Sequence-tagged site (STS) mapping:* It maps position of the short sequence by polymerase chain reaction (PCR).

Uses of Gene Mapping
- Identification of the gene responsible for heritable diseases or cancer.
- To establish a relationship between gene and the disease.
- To establish a relationship between the variation of the gene (SNP) and the disease, drug resistance or phenotypic changes.

Box 8: Human Genome Project

Human Genome Project

Q. Write a short note on Human Genome Project.
- It is a research program to locate the genes in the human genome and explore the details of genes.
- The human genome is a complete set of genetic information for humans.
- 1986: Human Genome Project started
- 2003: Human Genome Project sequencing completed.

Advantages of Human Genome Project
- Ability to locate genes that are responsible for locating diseases.
- Can be used for gene therapy.

Contd...

Structure of DNA and RNA   41

Contd...

Outcomes
- Human genome contains 22 autosomes and X- and Y-chromosomes.
- 6 feet DNA
- 30,000 genes
- 3 billion nucleotide pairs
- Average gene contain 3,000 bases.
- Only 3% genome encodes for proteins and rest of it junks DNA.
- Chromosome 1 has highest genes (2,968) and chromosome Y has the lowest (231).
- Mostly DNA of the individual differs from each other by SNPs.
- The approximate size of the diploid human genome is 3×10^9 base pairs (3 billion bp). *(NEET)*

GENE BANK

It is the collection of DNA molecules that possess complete genetic information of an organism.

Indian National Gene Bank
- National Bureau of Plant Genetic Resources (NBPGR) maintains Indian National Gene Bank.
- It preserves dehydrated plant seeds, tissue cultures, synthetic seeds, germplasms and so on.

Some Interesting Facts
- DNA polymerase and DNA ligase are essential enzymes for DNA synthesis.
- Topoisomerase unwinds DNA.
- Okazaki fragments are short, newly synthesised DNA fragments. They are synthesised on the lagging strand of DNA during its replication.
- DNA polymerase I completes DNA synthesis between Okazaki fragments on the lagging strand.
- The enzymes required for the formation of the DNA in the sequence are RNA polymerase, DNA polymerase III, DNA ligase, exonuclease and DNA polymerase I.
- During DNA replication (copying), most DNA polymerases (mostly polymerase II) can check their work (proofreading). Polymerases remove wrong (incorrectly paired) nucleotide and replace the nucleotide right away before continuing with DNA synthesis.
- Ultraviolet radiation induces the formation of cyclobutane pyrimidine-pyrimidine dimers.
- Xeroderma pigmentosa is the example of *pyrimidine dimer* formation and *mismatch* repair or low activity of excision repair process.
- Leucine zipper motif is a mediator for binding of the regulatory protein to DNA.
- Jacob and Monod elucidated an operon model. An operon is a functioning unit of genomic DNA containing a cluster of genes under the control of a single promoter.
- Zink finger is a nuclear receptor.

SOLUTION FOR CLINICAL CASE

Genetic counsellor should explain that this is a suspected case of Kearns-Sayre syndrome that shows mtDNA inheritance. As an individual receives mtDNA from the mother, this disease can be suspected in the mother also. Its diagnosis mostly depends on clinical analysis, changes on muscle biopsy (ragged-red fibers with the modified Gomori trichrome stain), immunohistochemical staining shows increased succinate dehydrogenase and absence of cytochrome C oxidase. Deletion of mtDNA is present in 90% of the cases that can be detected using PCR and RFLP.

CHAPTER 6

Laws of Inheritance

> **CLINICAL CASE**
>
> A man with brown eyes and curly, dark hairs was married to a woman with blue eyes and straight, light hairs. What will be the eye color and hair texture of their daughter and why?

INTRODUCTION

Mendelian Inheritance
- Gregor Mendel, Father of Modern Genetics, was born in 1822 in a gardener's family in Austria.
- He performed experiments in monastery gardens for 7 years and published his work at the National History Society of Brünn in 1965.
- His work remained neglected for 35 years.
- In 1900, Erich Von Tschermak, Hugo de Vries and Carl Correns rediscovered Mendel's principles.

Mendel's Work
- Mendel observed results of inheritance of seven different pairs of contrasting characters in *Pisum sativum* (garden pea).
- He used artificial pollination and cross-pollination for his experiments.

Reasons for Mendel's Success

Q. Enlist the reasons for Mendel's success.
- Selected *Pisum sativum* (Pea plant).
- Pea plant flowers are self-fertilised.
- Contrasting features of the pea plant.
- Easy cross-pollination.
- Easy cultivation of pea plants.
- Short growth period and short life cycle.
- Studied only one character at a time.
- Maintained proper experimental records.

CHARACTERS STUDIED BY MENDEL

The characters studied by Mendel are listed in **Table 1**.

Table 1: Pea plant characters selected by Mendel

Character	Variety	
	Dominant	Recessive
Stem length	Tall	Dwarf
Shape of seed	Round	Wrinkled
Seed color	Yellow	Green
Seed coat color	Gray	White
Position of flower	Axial	Terminal
Form of pod	Inflated	Constricted
Color of pod	Green	Yellow

MENDEL'S LAWS

Mendel stated three laws: Law of Dominance, Law of Incomplete Dominance, and Law of Segregation.

Law of Dominance

Q. Write a short note on Law of Dominance.
- Crossing between organisms for contrasting characters of a pair, only one character (dominant) appears in the first generation and another does not appear (recessive) **(Fig. 1)**.
- Every character is controlled by a pair of the gene.

Genotype
Genotype is the genetic constitution of an organism with respect to a character.

Phenotype
- The phenotype is the external appearance (or expressed protein) of an organism with respect to a character **(Table 2)**.

Table 2: Relation of phenotype and genotype

Phenotype	Genotype
Homozygous tall	TT
Heterozygous tall	Tt
Homozygous dwarf	tt

- Some examples of dominant and recessive characters in human are listed in **Table 3**.

Principles of Clinical Genetics

Tall pea plant TT → T, T (Gametes)
Dwarf pea plant tt → t, t (Gametes)

→ Tt
Tall pea plant

Selfing of first generation

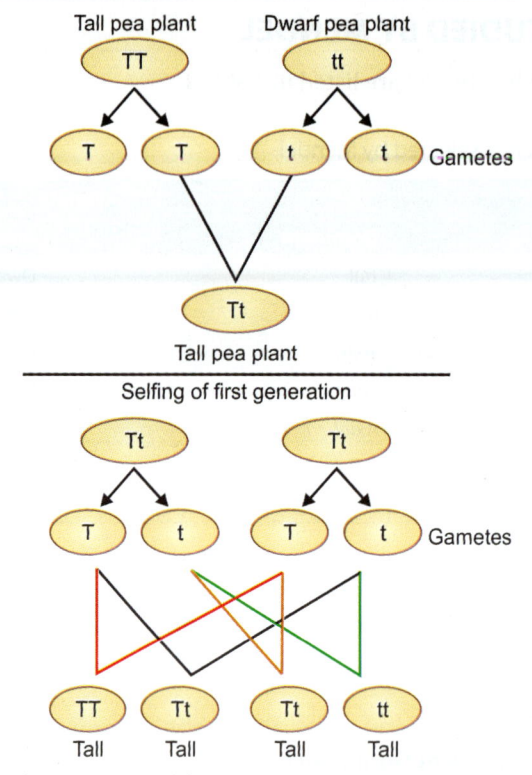

Tt → T, t (Gametes)
Tt → T, t (Gametes)

TT (Tall), Tt (Tall), Tt (Tall), tt (Tall)

Genotypic ratio: 1:2:1 and phenotypic ration: 3:1

Fig. 1: Law of dominance

Table 3: Dominant characters in human

Q. Enlist the dominant characters in human.

Character	Dominant	Recessive
Eye color	Brown	Blue
Hair	Curly, dark	Straight, light
Eyesight	Normal	Color blind
Ear lobe	Free	Attached

Law of Incomplete Dominance

Q. Write a short note on Law of Incomplete Dominance.
- Some characters are neither dominant nor recessive. These are expressed midway between two parents.
- For example, plant crossed with red flowers (RR) and white flowers (rr), first generation plant bears pink flowers (Rr).

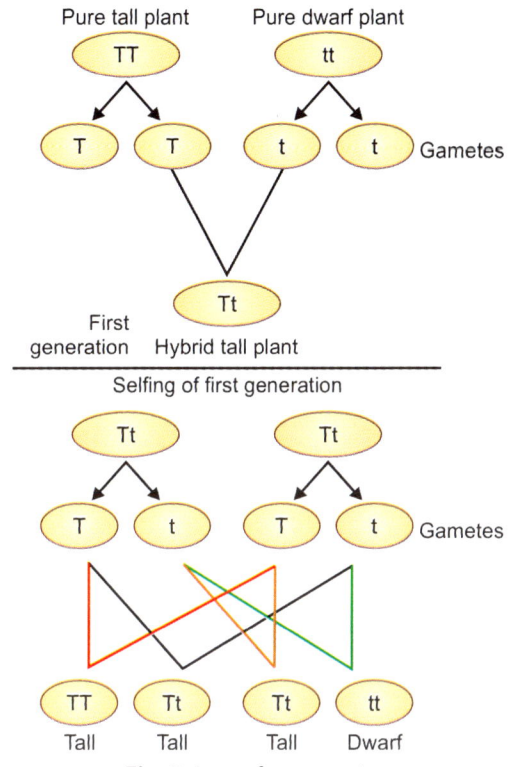

Fig. 2: Law of segregation

Law of Segregation

Q. Write a short note on Law of Segregation.

A pair of contrasting character in a hybrid form remains together without mixing with each other and segregates during formation of gametes **(Fig. 2)**.

Law of Independent Assortment

Q. Write a short note on Law of Independent Assortment.
- Two parents differing from each other in two or more pairs of contrasting characters inherit one pair of character independent of another pair of character.
- Law of independent assortment is proved by dihybrid cross **(Fig. 3)**.

Genotype: 9 different combinations

Phenotype: 9:3:3:1
9 – Yellow round
3 – Yellow wrinkled
3 – Green round
1 – Green wrinkled.

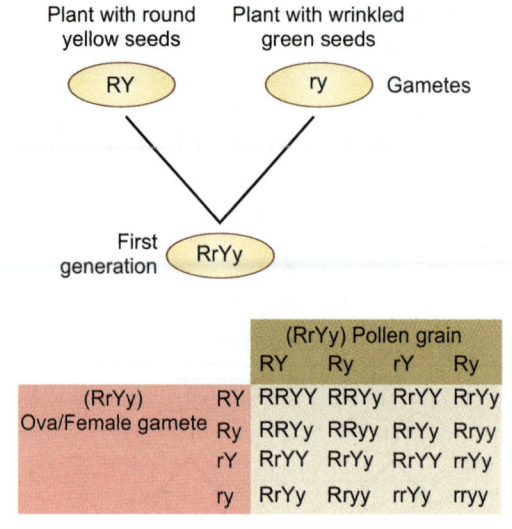

```
                    Plant with round      Plant with wrinkled
                     yellow seeds            green seeds

                        ( RY )                  ( ry )      Gametes



                         First      ( RrYy )
                       generation
```

		(RrYy) Pollen grain			
		RY	Ry	rY	ry
(RrYy)	RY	RRYY	RRYy	RrYY	RrYy
Ova/Female gamete	Ry	RRYy	RRyy	RrYy	Rryy
	rY	RrYY	RrYy	RrYY	rrYy
	ry	RrYy	Rryy	rrYy	rryy

Fig. 3: Law of independent assortment

BIOLOGICAL SIGNIFICANCE OF MENDEL'S LAWS

Q. Enlist the biological significances of Mendel's laws.
- Useful in the understanding inheritance of human diseases.
- Useful for improving life (positive eugenics) and preventing the passage of the diseases to next generations.
- For increasing agricultural production and other animal products.
- For eliminating recessive disorders with the help of genetic counseling.

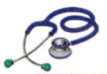

SOLUTION FOR CLINICAL CASE

Mendel's Law of Dominance: In human beings, brown eyes and curly, dark hairs are dominant phenotypic characters over blue eyes and straight, light hairs.

Mendel's Law of Segregation: A pair of contrasting character in a hybrid form remains together without mixing with each other and segregates during the formation of gametes. In the present case, the male may be heterozygous or homozygous for dominant characters but the female is surely homozygous for both the characters.

Mendel's Law of Independent Assortment: Though the eye and hair texture characters inherited together, they get separated at the time of gamete formation.

According to Mendel's laws, if the male partner is homozygous for both of the given characters, then the daughter will surely have brown eyes and curly, dark hairs. If the male partner is heterozygous for any one of the given characters, then daughter may have a chance of blue eye and straight, light hairs as per Mendel's laws.

CHAPTER 7

Patterns of Inheritance

CLINICAL CASE

Sickle cell trait (heterozygous state) patient do not have severe anemia such as a homozygous patient. But, sickle cell trait cases are not healthy individuals. Explain the genetic basis.

INTRODUCTION

Inheritance *(Viva)*
- All the characters of an individual are inherited from parents through genetic components.
- In genetic disorder, the expression of the character in an individual depends on the category of the disease.

Allele *(Viva)*
- An allele is a variant from a gene.
- Allele of the same gene is located at the same position (genetic locus) on the chromosome.
- In human, each gene has two alleles, one inherited from the mother and another from the father.

Homozygous and Heterozygous *(Viva)*
- An individual is called homozygous for the gene if he has two identical copies of the same allele for that gene (for example, AA or aa).
- If an individual has nonidentical copies of the same allele (for example, Aa), the individual is called heterozygous for that gene.

Dominant and Recessive *(Viva)*
- Alleles can be either dominant or recessive.
- Each allele may have a phenotype (expressed protein or external character).
- In heterozygous condition, if one allele is expressed and another allele is masked, then the expressed allele is called dominant and the masked allele is called recessive.
- Codominant is the condition where both the alleles are expressed simultaneously to generate a mixed phenotypic result, for example, blood group alleles.

Multiple Alleles
- Most of the genes exist in the form of two alleles.
- Some genes have multiple alleles for a particular character. For example, ABO blood group type.

Polygenic Trait
- A polygenic trait is a character that is determined by more than one gene.
- Examples: height, weight, skin color.

Phenotype
The phenotype is an outward physical characteristic manifestation of the organism. It also includes expressed proteins, macromolecules, cell organelles, cell structure or function and behavior of the organism.

Genotype (Complete Set of Genes)
Genotype is the total genetic makeup of an individual that determines the phenotype.

Genetic Disorder
- A genetic disorder is a condition caused due to an abnormality in the genome of an individual (**Box 1**).

> **Box 1:** Importance of inheritance pattern analysis
> Study of inheritance patterns in the genetic disorders is very important for:
> - Assessing the status of the condition.
> - Diagnosing the risk of transmission.
> - Calculating the risk of transmission.
> - Planning for the treatment and prevention of the condition to next generation.

- There are three basic genetic disorders:
    1. Single gene disorders (Mendelian disorders)—defect in a single gene.
    2. Chromosomal disorders—defect in the number or structure of chromosomes.
    3. Multifactorial disorders—defective interaction between genes and environmental factors (viruses, drugs, food, etc.).

PEDIGREE
Q. Write a short note on pedigree.

Definition
- A pedigree is a method for documenting the family relationship of an individual under study.
- Symbols of pedigree are given in **Figures 1 and 2**.

Advantages of Pedigree Charting
- Provides family information in an easily readable chart.
- Provides information on the mode of inheritance.
- Provides assistance in planning for treatment and prevention of transmission of a genetic disorder to next generation.
- Helps for genetic counseling.

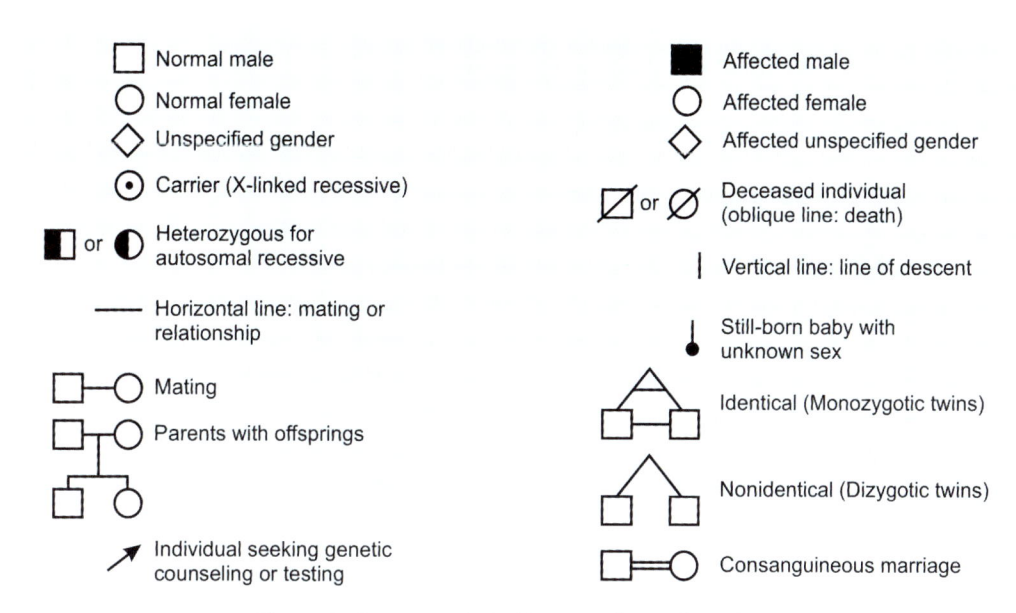

Fig. 1: Pedigree symbols (Commonly used symbols)

For details, read Benett RL, French KS, Resta RG, Doyle DL. Standardized human pedigree nomenclature: update and assessment of the recommendations of the National Society of Genetic Counselors. J Genet Couns. 2008;17:424-33.

SINGLE GENE INHERITANCE

- Mutation in single gene → monogenic genetic disorder → shows single gene inheritance or Mendelian inheritance.
- Single gene disorders are grouped as follows:
  - Autosomal dominant
  - Autosomal recessive
  - X-linked dominant
  - X-linked recessive
  - Y-linked disorders.
- As Y-linked disorders are carried only from father to son, sex-linked disorder term is confined for X-linked disorders.

Autosomal Dominant Inheritance

Q. Write a short note on autosomal dominant inheritance.

Definition
- In autosomal dominant disorder, even if a single abnormal gene (allele) is present, it is expressed phenotypically and produces a disorder **(Figs 3 and 4) (Table 1)**.
- For autosomal dominant state:
  - Heterozygotic state → mild disease
  - Homozygotic state → severe disease (due to both abnormal genes).

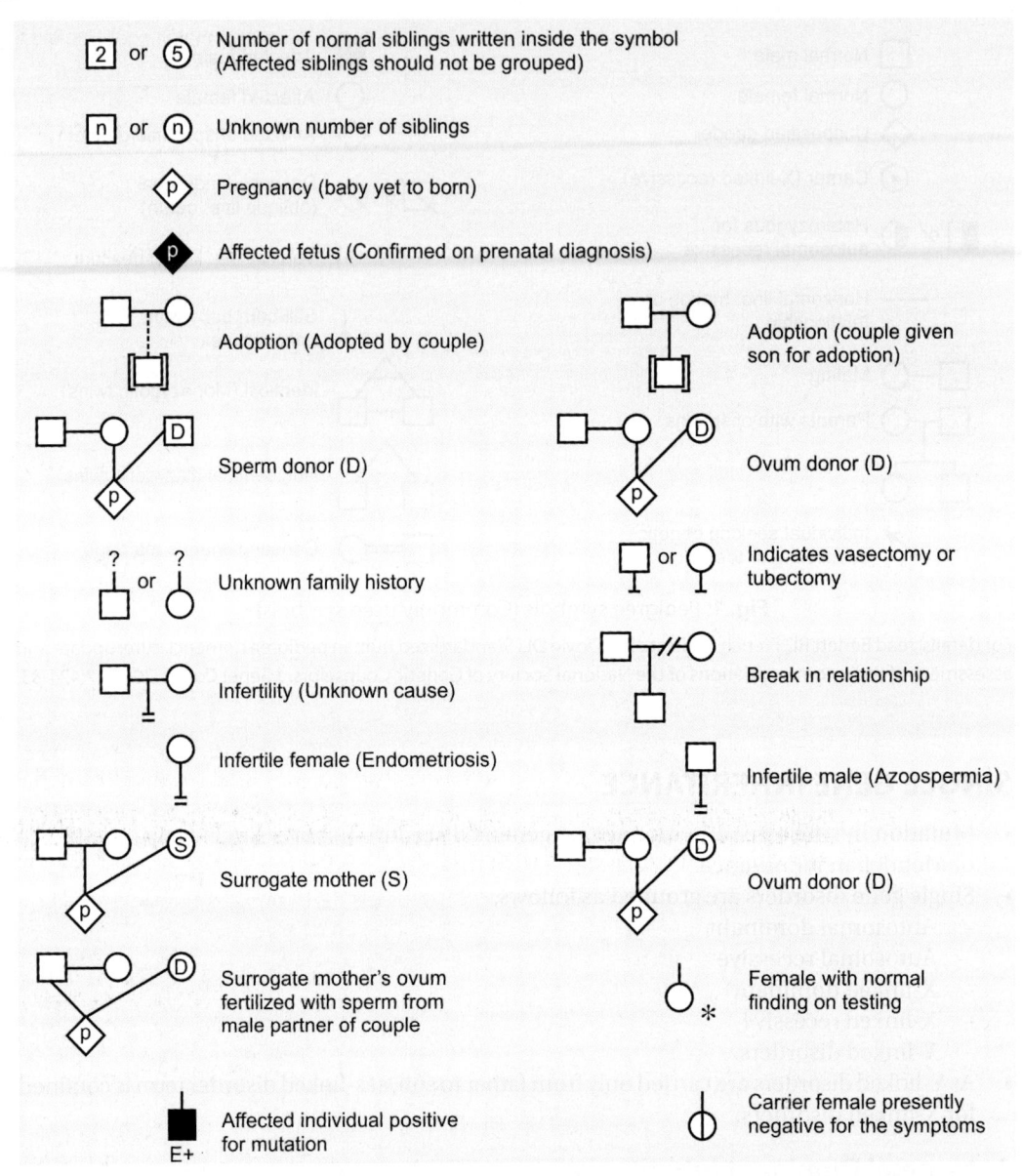

Fig. 2: Pedigree symbols (symbols for special cases)

Common Features
- Affects irrespective of sex of the individual
- Affects all generations (vertical pattern of transmission)
- 50% offspring of affected individuals will be normal and 50% will be affected.
- Normal child is not carrier (do not transmit the disease).
- Affected gene may be transmitted from or by new mutation during gametogenesis.
- Variable clinical expression

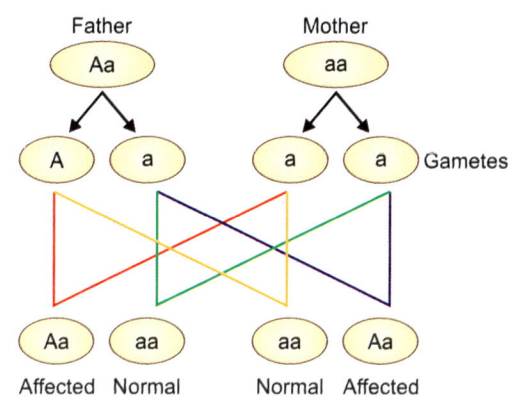

Fig. 3: Autosomal dominant inheritance

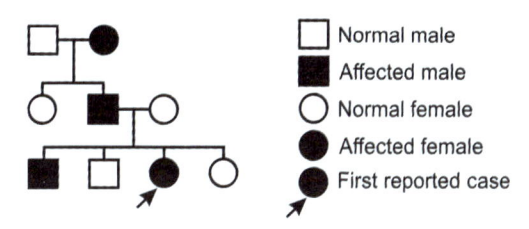

Fig. 4: Pedigree for autosomal dominant inheritance

Table 1: Autosomal dominant disorders *(NEET)*

	Disease	Affected gene
1.	Achondroplasia	Fibroblast growth factor receptor 3
2.	Hypercholesterolemia	LDL receptor
3.	Holoprosencephaly	Sonic Hedgehog (*SHH*)
4.	Huntington chorea	Huntingtin (HD)
5.	Marfan syndrome	Fibrin-1 gene
6.	Myotonic dystrophy	A protein kinase gene (*DMPK*)
7.	Neurofibromatosis 1	Microdeletion at 17q 11.2 involves *NF1* gene
8.	Osteogenesis imperfecta	Genes encoding α1 or α2 chain of type-1 collagen
9.	Polycystic kidney disease	Polycystin-1 or polycystin-2 gene
10.	Waardenburg syndrome	*PAX3* gene located at 2q35
11.	Tuberous sclerosis/Bourneville-Pringle disease	Tuberous sclerosis (*TSC1* and *TSC2*) genes
12.	Treacher Collins syndrome/mandibulofacial dysostosis	*TCOF-1* gene, POLRIC or POLRID genes (all are involved in pharyngeal arch development)

Abbreviation: LDL, low-density lipoprotein

- Autosomal dominant inheritance is not related to consanguineous marriage.
- Affected individual may be homozygous or heterozygous.

Autosomal Recessive Inheritance

Q. Write a short note on autosomal recessive inheritance.

Definition
- In autosomal recessive disorder, if both the copies of allelic genes are abnormal, then only a disease is expressed **(Figs 5 and 6) (Table 2)**.

Fig. 5: Autosomal recessive inheritance

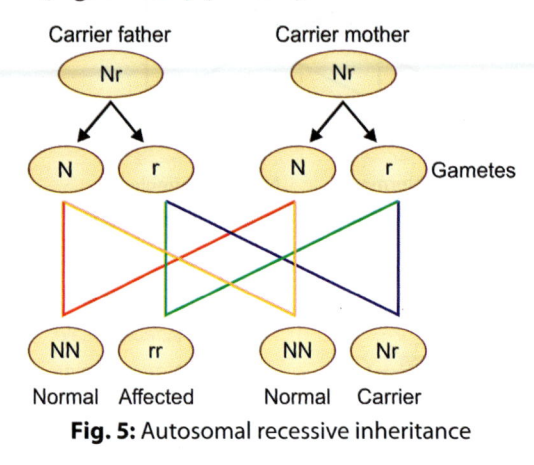

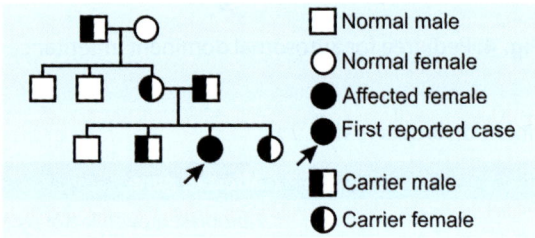

Fig. 6: Pedigree for autosomal recessive inheritance

Table 2: Autosomal recessive disorders (*NEET*)

	Disorder	Affected gene
1.	Cystic fibrosis	Cystic fibrosis transmembrane regulator (CFTR) → impaired chloride ion channel function
2.	Gaucher's disease (lysosomal storage disorder)	β-glucosidase
3.	Hemochromatosis (enhanced iron absorption and abnormal deposition)	Unknown gene on the short arm of chromosome 6
4.	Phenylketonuria	Phenylalanine hydroxylase
5.	Tay-Sachs disease	β-hexosaminidase
6.	Xeroderma pigmentosa	Genes involved in nucleotide excision repair
7.	Sickle cell anemia	Hemoglobin beta gene on chromosome 11
8.	Alkaptonuria	Homogentisate 1,2-dioxygenase gene
9.	Albinism	Tyrosinase-related protein 1 gene
10.	Spinal muscular dystrophy	Survival of motor neuron 1 (*SMN1*) gene

- For autosomal recessive state:
  - Heterozygotic → no disease
  - Homozygotic → disease.

Common Features
- Male and female are equally affected
- *Prediction:* 25% children will be affected (risk is 1 in 4), 25% will be normal and 50% will be carrier.
- The unaffected child will not transmit the disease.
- May be a result of consanguineous marriage.
- Early age of onset.
- Constant clinical features are seen.

The differences between autosomal dominant and recessive inheritance are given in **Table 3**.

X-linked Dominant Inheritance

Q. Write a short note on X-linked dominant inheritance.

Definition
X-linked dominant inheritance is a pattern of inheritance that involves mutated dominant gene of X-chromosomes **(Figs 7 and 8) (Table 4)**.

Common Features
- Only one affected gene can produce disease.
- No male preponderance such as X-linked recessive disorders
- All the daughters of affected male will also be affected and all sons of affected mother will also be affected.
- Males have only one X-chromosome, that is, hemizygous males always show aggravated disease.

Table 3: Comparison of autosomal dominant and autosomal recessive inheritance

Autosomal dominant inheritance	Autosomal recessive inheritance
In autosomal dominant disorder, even if a single allelic gene is present, it is expresses and produces disorder	In autosomal recessive disorder, only if both the copies of allelic genes are abnormal then only the disease is produced.
Heterozygous individual—affected	Heterozygous individual—normal
Homozygous individual—affected	Homozygous individual—affected
An unaffected individual cannot transmit the disease	An unaffected individual can transmit the disease if he/she is a carrier.
No carrier state	Carrier state present.
Known as hereditary diseases	Known as familial diseases.
No relation with consanguineous marriages	Consanguineous marriages increase the risk of disease.
Examples: Achondroplasia, Marfan syndrome.	*Examples:* Cystic fibrosis, phenylketonuria

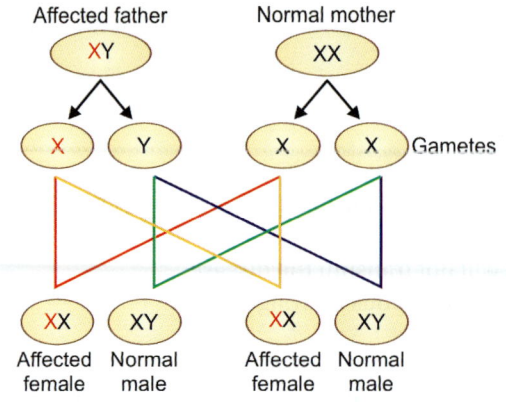

Fig. 7: X-linked dominant inheritance

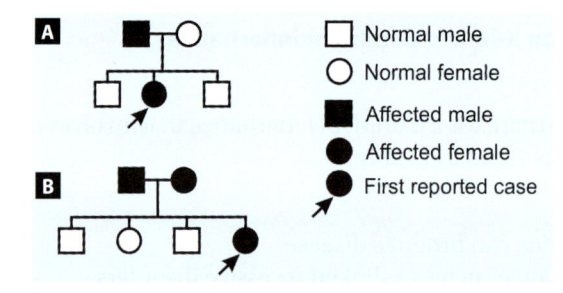

Fig. 8: Pedigree for X-linked dominant inheritance (A and B are separate case reports)

Table 4: X-linked dominant inheritance *(NEET)*

	Syndrome	Affected gene
1.	Fragile X-syndrome	Fragile X-mental retardation gene (*FMR*)
2.	X-linked hypophosphatemia (vitamin D-resistant rickets)	Phosphate-regulating neutral endopeptidase (*PHEX* gene)
3.	Rett syndrome (cerebroatrophic hyperammonemia)	Methyl-CpG binding protein 2 (*MECP2*) gene
4.	Alport syndrome	Collagen-alpha-5 (IV) chain [*COL4A5*] gene
5.	Focal dermal hypoplasia or Goltz syndrome	Porcupine homolog drosophila (*PORCN*) gene
6.	Xg blood group system: This is normal blood grouping of any disorder.	*XG* gene

X-linked Recessive Inheritance

Q. Write a short note on X-linked recessive inheritance.

Definition

X-linked recessive inheritance is a pattern of inheritance that involves mutated recessive gene of X-chromosome **(Figs 9 and 10) (Table 5)**.

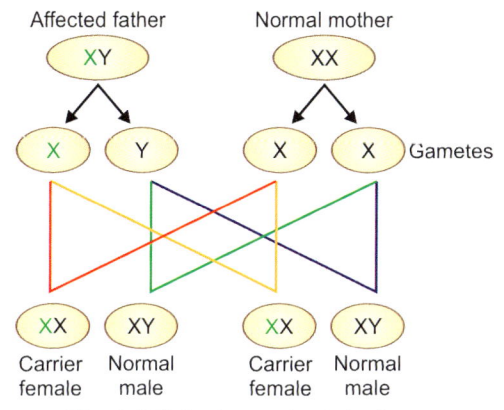

Fig. 9: X-linked recessive inheritance

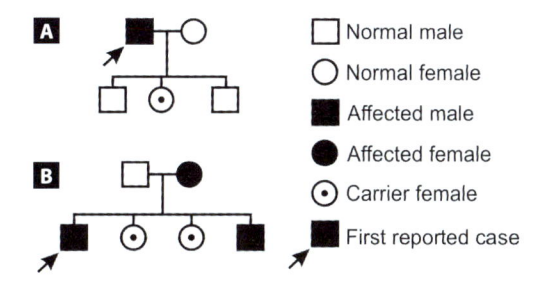

Fig. 10: Pedigree for X-linked recessive inheritance (A and B are separate case reports)

Expression
- Homozygous female → affected case.
- Heterozygous female → carrier.
- Hemizygous male (because the male has only X-chromosome) → affected case.

Common Characteristics
- The heterozygous female is a carrier and the homozygous female is affected case.
- As males carry only one X-chromosome (hence, hemizygous), they have the disease.
- Sons of the affected mother are always affected.
- Daughter of the affected father is a carrier, if the mother is normal.
- Sons of the affected father are normal.
- More common in males.

Y-linked Inheritance/Holandric Inheritance

Q. **Write a short note on Y-linked inheritance.**

Q. **Write a short note on holandric gene.**

Definition
The pattern of inheritance of mutated Y-chromosome gene from father to son is called Y-linked inheritance.

Table 5: X-linked recessive inheritance *(NEET)*

	Disorder	Affected gene
1.	Duchenne muscular dystrophy	Dystrophin
2.	Glucose 6-phosphate dehydrogenase deficiency	Glucose 6-phospahate dehydrogenase
3.	Hemophilia A	Factor VIII gene
4.	Hemophilia B	Factor IX gene
5.	Becker's muscular dystrophy	Dystrophin
6.	X-linked ichthyosis	Steroid sulfatase enzyme gene
7.	X-linked agammaglobulinemia or Bruton syndrome	Tyrosine kinase
8.	Red-green color blindness	• Opsin 1, long wave sensitive (OPN1LW) for red cone pigment • Opsin 1, medium wave sensitive gene (OPN1LW) for green cone pigment
9.	Barth syndrome or 3-methylglutaconic aciduria type-II	Tafazzin gene (acyltransferase)
10.	Lesch-Nyhan syndrome or juvenile gout (hyperuricemia)	Hypoxanthine-guanine phosphoribosyltransferase (HGPRT) gene
11.	Norrie disease (cataract, leukocoria, blindness)	Norrie disease protein (NDP gene)

Common Features
- Affects only the male.
- All sons of affected males are also affected.
- Females are neither affected nor a carrier.

Examples
- Hairy pinna
- Baldness.

POLYGENIC INHERITANCE (MULTIFUNCTIONAL INHERITANCE)

Q. **Write a short note on polygenic or multifactorial inheritance.**

Definition

In polygenic inheritance, a group of genes interact with each other to produce a disease **(Box 2)**.

Characteristics
- Polygenic inheritance does not follow Mendelian inheritance.
- It depends on multiple genes.
- It also depends on environmental factors.

- Variable expression in all individuals, family members and even in monozygotic twins.
- Polygenic genes tend to aggregate in families.
- The incidence of the condition is greater in relatives of the affected individual than that in general population.
- The risk is greatest for the first-degree relatives.
- Consanguinity increases the risk.

> **Box 2:** Disease with polygenic inheritance
>
> Study of inheritance patterns in the genetic disorders is very important for:
> A. Congenital malformations
>   - Cleft lip, cleft palate
>   - Congenital heart defects
>   - Neural tube defects
>   - Pyloric stenosis
>   - Talipes
> B. Adult onset diseases
>   - Diabetes mellitus
>   - Epilepsy
>   - Glaucoma
>   - Hypertension
>   - Depression
>   - Schizophrenia
>   - Asthma
>   - Spondylitis

Some Interesting Facts

Codominant Inheritance
- Codominant genes are expressed simultaneously to produce a phenotype.
- *Example:* A gene and B gene are expressed simultaneously to produce AB group. *(NEET)*

Intermediate Inheritance
- Some genes in the heterozygous state also get expressed to a certain extent but not like dominant genes.
- *Example:* Sickle cell trait (heterozygous state) patient do not have severe anemia such as a homozygous patient. But trait cases are not normal individuals. Thus, expression of sickle cell trait is intermediate inheritance. *(NEET)*

Sex-limited Traits
- Sex-limited genes are present in both the sexes but expressed only in one sex and in another remain "turned off". These gene traits are called sex-limited traits.
- *Example:*
    - Development of facial hairs in the male.
    - Development of breast in females.

Sex-influenced Traits
- Some genes inherit as autosomal traits that get influenced by sex.
- *Example:* Pattern of baldness is influenced by testosterone.

Principles of Clinical Genetics

Box 3: Mitochondrial inheritance

Mitochondrial inheritance

Q. Write a short note on mitochondrial inheritance.
- A mitochondrion contains deoxyribonucleic acid (DNA) called mitochondrial DNA (mtDNA) **(Fig. 11)**.
- Mitochondrion in human are contributed from the mother (cytoplasm of ovum), therefore inheritance of mitochondrial DNA is also called maternal inheritance.
- Human mitochondrial DNA codes for 37 genes.

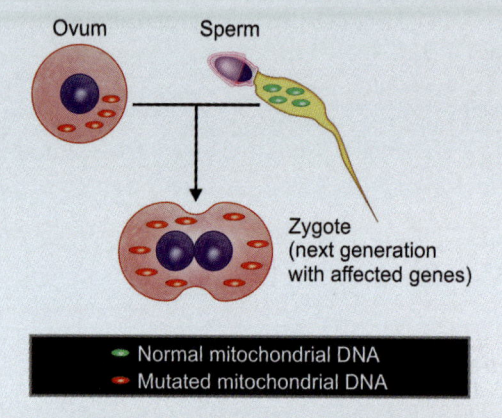

Fig. 11: Mitochondrial inheritance
Abbreviation: DNA, deoxyribonucleic acid

- Examples of mitochondrial inheritance *(NEET)*
    - Mutations in cytochrome b gene in mitochondrial DNA → Exercise intolerance.
    - Mitochondrial DNA deletion → Kearns-Sayre syndrome (mitochondrial myopathy).
    - Mitochondrial DNA deletion → Pearson syndrome (sideroblastic anemia and exocrine pancreas dysfunction).
    - Mutated mitochondrial DNA → Ragged red fibers.
    - Mutated mitochondrial DNA → Leber's hereditary optic neuropathy → (retinal ganglionic cell degeneration).

Box 4: Pleiotropy

Pleiotropy

Definition

This condition occurs when one gene influences many phenotypic traits.

Example
- *Phenylketonuria:* In this condition, enzyme phenylalanine hydroxylase is deficient. In addition to phenylketonuria, the child develops hypopigmentation and mental retardation. *(NEET)*
- *Galactosemia:* In this condition, galactose-1-phosphate uridyltransferase deficiency → galactosemia with liver cirrhosis, cataract and mental retardation. *(NEET)*

Box 5: Incomplete penetrance *(NEET)*

Incomplete penetrance
- Penetrance is the proportion of individuals carrying mutated gene with clinical symptoms.
- Incomplete penetrance indicates nonexpression of the mutated gene due to the influence of other genes.

For example:
- Breast cancer gene 1 (*BRCA*1) mutation produces breast cancer only in some females.
- *Polydactyly gene:* Some individuals have incomplete extra finger and some have a complete extra finger.

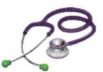

SOLUTION FOR CLINICAL CASE

Intermediate inheritance: Some genes in the heterozygous state also get expressed to a certain extent but not like dominant genes. Expression of sickle cell trait is intermediate inheritance. Hence, in sickle cell trait (heterozygous state), patient do not have severe anemia compared to a homozygous patient. Thus, trait cases are not healthy individuals.

CHAPTER 8

Inborn Errors of Metabolism

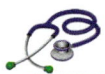

CLINICAL CASE

A 2-year-old child has intellectual developmental delay, gradual change of hair color from black to light brown, positive Guthrie test and plasma amino acid analysis showing a phenylalanine level of 15 mg/dL (normal levels are 61–121 μmol/L or 1–2 mg/dL), urine neopterin to biopterin ratio and blood dihydropteridine reductase activity are normal. What may be the probable diagnosis, its cause and treatment of choice?

INTRODUCTION

Definition
- Inborn errors of metabolism are inherited disorders in which body cannot metabolise nutrients.
- Inborn errors of metabolism is produced due to defect in the gene coding an enzyme that is essential for the conversion of substrate into product.
- Archibald Garrod (1908) coined the term inborn errors of metabolism.

PATHOGENESIS OF INBORN ERRORS OF METABOLISM

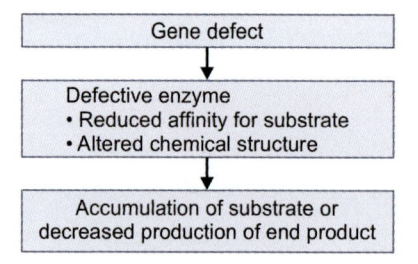

```
           Gene defect
                ↓
         Defective enzyme
      • Reduced affinity for substrate
      • Altered chemical structure
                ↓
     Accumulation of substrate or
    decreased production of end product
```

- *Note:* Abnormal cell surface receptors results in a nonmetabolic error, for example, defective low-density lipoprotein (LDL) receptors → familial hypercholesterolemia.
- Classification of inborn errors of metabolism is given in **Table 1**.

Table 1: Inborn errors of metabolism

Group	Mode of inheritance	Enzyme
Disorders of carbohydrate metabolism		
Glycogen storage diseases		
von Gierke's disease (GSD-I)	AR	Glucose 6-phosphatase *(NEET)*
Pompe's disease (GSD-II)	AR	Acid alpha-glucosidase
Cori's disease (GSD-III)	AR	Glycogen debranching enzyme *(NEET)*
Andersen disease (GSD-IV)	AR	Glycogen-branching enzyme
McArdle disease (GSD-V)	AR	Muscle glycogen phosphorylase
Hers disease (GSD-VI)	AR	Liver glycogen phosphorylase
Tarui's disease (GSD-VII)	AR	Muscle phosphofructokinase
Favism (G6PD)	XR	Glucose-6-phosphate dehydrogenase
Disease of amino acid metabolism		
Phenylketonuria	AR	Phenylalanine hydroxylase
Maple syrup urine disease	AR	Branched-chain-alpha-keto acid dehydrogenase
Glutaric aciduria type I	AR	Glutaryl-CoA dehydrogenase
Albinism	AR	Tyrosinase *(NEET)*
Homocystinuria	AR	Cystathionine β-synthase 1
Disorders of urea cycle		
Carbamoyl phosphate synthetase I deficiency	AR	Carbamoyl phosphate synthetase 1
Disorders of organic acid metabolism		
Alkaptonuria	AR	Homogentisic acid oxidase *(NEET)*
Disorders of fatty acid oxidation		
Medium-chain acyl-coenzyme A dehydrogenase deficiency (MCADD)	AR	Acyl-CoA dehydrogenase deficiency
Disorders of porphyrin metabolism		
Acute intermittent porphyria	AD	Porphobilinogen deaminase
Disorders of purine/pyrimidine metabolism		
Lesch-Nyhan syndrome (Juvenile gout)	XR	Hypoxanthine-guanine phosphoribosyl transferase (HGPRT) *(NEET)*
Disorders of steroid metabolism		
Congenital adrenal hyperplasia	AR	21-hydroxylase *(NEET)*
Disorders of mitochondrial functions		
Kearns-Sayre syndrome	mtDNA	Mutation in mitochondrial DNA

Contd...

Contd...

Group	Mode of inheritance	Enzyme
Disorders of peroxisomal function		
Gaucher's disease	AR	Glucocerebrosidase *(NEET)*
Niemann-Pick disease	AR	Lysosomal enzyme acid sphingomyelinase
Classic galactosemia	AR	Galactose-1-phosphate uridyl transferase
Hurler syndrome (mucopolysaccharides I)	AR	α-iduronidase
Hunter syndrome (mucopolysaccharides II)	XR	Iduronate sulfatase
Sly syndrome	AR	β-glucuronidase
Maroteaux–Lamy syndrome	AR	Arylsulfatase B
Tay–Sachs disease	AR	Hexosaminidase A deficiency *(NEET)*
Disorders of copper metabolism		
Wilson disease	AR	Copper metabolism *(NEET)*
Menkes disease	XR	ATPase membrane copper transporter protein

Abbreviations: AR, autosomal recessive; AD, autosomal dominant; XR, X-linked recessive

PHENYLKETONURIA

Q. Write a short note on phenylketonuria.
- First discovered by Ivar Asbjørn Følling (1934).
- It is an inborn error of metabolism.

Cause
Defective or absence of phenylalanine hydroxylase.

Inheritance
Autosomal recessive.

Role of Enzyme

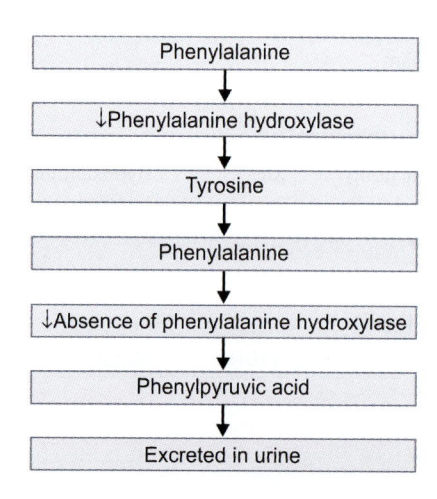

Phenylalanine
↓
↓Phenylalanine hydroxylase
↓
Tyrosine
↓
Phenylalanine
↓
↓Absence of phenylalanine hydroxylase
↓
Phenylpyruvic acid
↓
Excreted in urine

- *PAH gene location:* Chromosome 12 in 12q22-q24.1.

Clinical Features
- Intellectual instability, mental disorders.
- Seizures
- Tyrosine deficiency → blonde hairs, blue eyes, lighter skin.

Diagnosis
Early diagnosis is essential, as brain changes are irreversible.
- *Guthrie test/neonatal heel prick:* It is a bacterial inhibition assay for detecting high serum phenylalanine level. *(NEET)*
- Tandem mass spectrometry for measuring the concentration of phenylalanine.

Treatment
Dietary control over phenylalanine

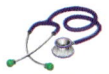

SOLUTION FOR CLINICAL CASE

The absence of phenylalanine hydroxylase produces phenylketonuria (autosomal recessive) disease. Phenylalanine hydroxylase deficiency causes decreased production of tyrosine, DOPA and melanin that result in light skin and hair color and an arrested intellectual development. Increased plasma phenylalanine is converted to phenylpyruvic acid and is excreted in urine. Treatment of choice is phenylalanine-restricted diet.

CHAPTER 9

Dermatoglyphics

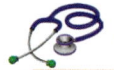

CLINICAL CASE

In the labor room, the nurse has taken prints of fingers, palm and sole of all babies in corresponding case sheets properly and each baby was marked with a hospital case sheet number tag. On the second day, two male babies were found with the missing tag. Their mothers were unable to identify the babies. The nurse asked duty doctor about the utility of prints of fingers, palm and sole for identification of babies. How should the doctor respond?

INTRODUCTION

Q. Write a short note on dermatoglyphics.

Definition

Dermatoglyphics (In Greek, derma = skin, glyph = carving) is the scientific study of patterns of dermal ridges on the palmar surface of digits, palm and sole **(Box 1)**.

Some Interesting Facts
- Sir Francis Galton reported observations on fingerprints in 1888.
- Dr Harold Cummins (1926) coined the term "dermatoglyphics".
- National Academy of Science, United States of America, in 2009, questioned for utilisation of dermatoglyphics in clinical diagnosis.

Box 1: Development of dermal ridges

The specific pattern of dermal ridges begins to develop during 13th week and is completed by the 19th week of intrauterine life.

Applications of Dermatoglyphics
- Clinical diagnosis of chromosomal anomalies such as Down's syndrome.
- Prediction of certain diseases such as essential hypertension.

- Forensic utility for identification of criminals.
- Record keeping and biometric identification.

ANATOMY OF FINGERPRINT

- The skin consists of two layers—epidermis (outer layer) and dermis (inner layer).
- The dermis has small projections that extend in the zone of the epidermis and creates papillary ridges.
- On the epidermal (skin) surface, these ridges form a specific pattern called fingerprint.

PATTERNS IN DERMATOGLYPHICS

The patterns of creases and ridges studied in dermatoglyphics are listed in **Box 2**.

Box 2: Patterns in dermatoglyphics

1. Flexor creases of palm
2. Dermal patterns
   a. Finger patterns
      - Simple arch
      - Loop
      - Whorl
   b. Palmar patterns
   c. Plantar patterns

Flexion Creases of Palm

- Palmar surface of hand shows palmar creases.
- They are as follows **(Fig. 1)**:
  - Wrist creases
  - Thenar crease
  - Proximal palmar crease
  - Distal palmar crease
  - Palmar digital crease **(Box 3)**
  - Proximal and distal interphalangeal creases on all fingers except thumb (thumb has only one interphalangeal crease).

Box 3: Palmar creases

- Simian crease—a single transverse crease across the palm instead of proximal and distal palmar creases. It is a characteristic feature of Down's syndrome. It is normally found in 1.5% of a general population at least in one hand. *(NEET)*
- Sydney line—proximal transverse crease crosses the palm completely.
- Suwon crease—proximal and distal palmar creases fuse with each other and there is an additional accessory proximal palmar crease **(Fig. 2)**.

Finger Patterns

The dermal papillae produce specific finger pattern **(Fig. 3)**.

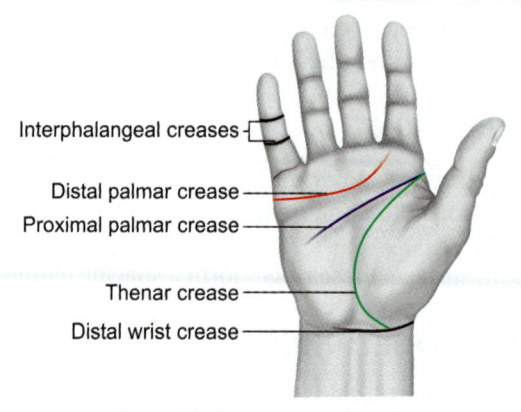

Fig. 1: Flexion creases of palm

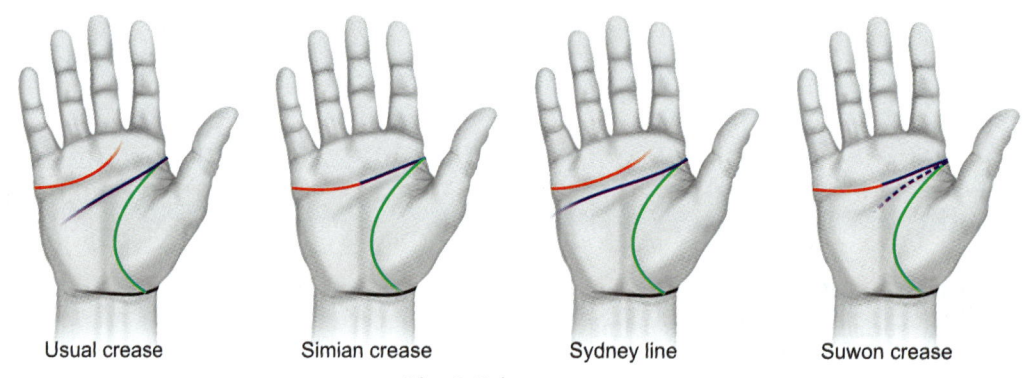

Usual crease   Simian crease   Sydney line   Suwon crease

Fig. 2: Palmar creases

Arch   Loop   Whorls

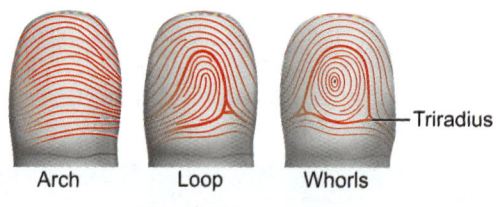

Fig. 3: Ridge patterns

GALTON CLASSIFICATION OF FINGER PATTERNS

- Galton classification of finger patterns based on triradii.
- Triradius is a point formed when three ridges course in three different directions at an angle of about 120°.
- There are three classes of finger ridges:
  1. Simple arch—no triradius
  2. Loop—one triradius
  3. Whorl—two or more triradii.

- Loops are of two types—ulnar (tail end of loop towards ulnar side of palm) and radial (tail end of loop towards radial side of palm)

Ridge Count
- It represents size of ridge count.
- *Method:* Counting of total number of ridges lying across the line drawn from the center of triradius to the center of finger pattern.
- For simple arch, ridge count is zero. *(NEET)*

Total Ridge Count
- Total ridge count (TRC) is sum of ridge count in all fingers (Both hands together).
- Increased TRC in Turner syndrome. *(NEET)*
- Decreased TRC in Klinefelter syndrome. *(NEET)*

Some Interesting Facts
- Most common finger pattern—loop.
- Greater the number of sex chromosomes, lower the total ridge count.

"atd" Angle
It is an angle formed by joining most proximal triradius on the hypothenar eminence of palm and the triradii below index and little finger **(Fig. 4)**.

"atd" Angle Significance
- < 35°—born athlete
- 35-45°—normal people
- 46-50°—slow learner
- 50°—mentally retarded
- In Down syndrome, usually atd angle is greater than 75°.

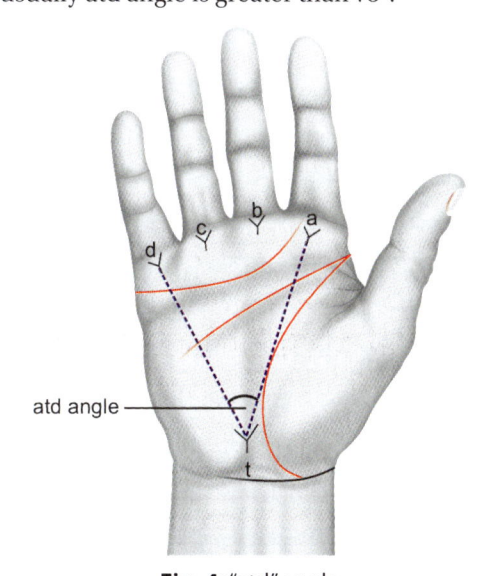

Fig. 4: "atd" angle

SIGNIFICANCE OF DERMATOGLYPHICS

Q. Write a short note on significance of dermatoglyphics.

Following are the significant applications:
- Anthropological applications of dermatoglyphics
- Inheritance pattern of different dermatoglyphics traits
- Dermatoglyphics in personal identification
- Association of dermatoglyphics pattern and diseases
- Dermatoglyphics in disputed paternity
- Dermatoglyphics in sports selection
- Association of dermatoglyphic patterns and intelligence of individual.

SOLUTION FOR CLINICAL CASE

The doctor should explain the mothers and nurse that dermatoglyphics is useful for the identification of the individual. The impressions of finger, palm and sole are unique for each individual. These impressions remain same throughout the life. Hence, babies can be identified easily using dermatoglyphics provided, it has been recorded properly at the time of birth of these babies.

SUGGESTED READING

1. Dorjee B, Mondal N, Sen J. Applications of dermatoglypics in anthropological research: a review. South Asian Anthropol. 2014;14(2):171-80.

CHAPTER 10

Cancer Genetics

INTRODUCTION
- Cancer is a disease of uncontrolled cellular proliferation on the transformation of normal cells.
- Cancerous cells may invade adjacent structures or metastasise (spread) to the distant body part.
- It also involves suppression of apoptosis (programmed cell death).

CANCER-CAUSING GENES

Q. Enlist the genes causing cancer.
Cancer causing genes are:
- Oncogenes
- Tumor suppressor genes
- DNA repair genes
- *TP53* genes
- Genes controlling cell death.

Oncogenes

Q. Write a short note on oncogenes.
- Oncogenes are the genes that get activated to produce tumor.
- Proto-oncogenes are normal genes that on mutation produce oncogenes.
- Oncogenes overcome the apoptosis and allow the cells to proliferate.
- Examples of oncogenes are given in **Table 1**.

Tumor Suppressor Genes

Q. Write a short note on tumor suppressor genes.
- It is also known as *antioncogene* (**Table 2**).
- It *protects the cell from cancer* by keeping break on undue cell proliferation.

Functions of Tumor Suppressor Genes
- Repression of genes that are required for cell growth and proliferation.
- Stops division of cells that get damaged DNA.

Table 1: Oncogenes and associated cancers

Oncogene	Chromosome location	Function	Associated tumors
Growth factor genes			
c-sis	22q12	Induces cell proliferation	Glioblastoma, melanomas
Growth factor receptor genes			
Epidermal growth factor receptor (ERBB)	7q12	Cell growth and differentiation	Breast cancer
Receptor tyrosine kinase (RET)	10q	Cell growth and differentiation	Thyroid cancer, multiple endocrine neoplasia
Signal transduction genes			
GTPase (KRAs)	12q14	Signaling for cell proliferation	Colon cancers, thyroid carcinoma, melanoma
Cyclin-dependent kinase	–	Cell cycle regulation	Malignant melanoma

Table 2: Tumor suppressor genes and associated tumors

Gene	Type of cancer	Function	Chromosome location
p53	More than 50% of all human cancers	Regulates cell cycle and apoptosis	17q13
p16	Melanoma, oropharyngeal squamous cell carcinoma, cervical cancers	Cyclin-dependent kinase inhibitor	9p21
APC	Familial adenomatous polyposis (colorectal cancer)	Cell adhesion	5q21
BRCA1	Hereditary breast cancer, ovarian cancers	Repair damaged DNA or induces apoptosis	17q21
BRCA2	Hereditary breast cancer, ovarian cancers	Repair damaged DNA or induces apoptosis	13q12
NF1	Neurofibromatosis type 1	Inhibits RAS transduction pathway	17q11
NF2	Neurofibromatosis type 2	Production of merlin or schwannomin protein for cell adhesion	22q12
RB1	Retinoblastoma, osteosarcoma	Inhibit cell cycle	13q14

- Some tumor suppressor genes repair damaged DNA and called *caretaker genes*, for example, BRCA1, BRCA2.
- Some tumor suppressor genes induce apoptosis (cell death) in cells with damaged DNA and such genes are called *gatekeeper genes,* for example, *p53, p21* genes.
- Increases cell adhesion and prevents metastasis.

TP53 GENE

Q. Write a short note on *TP53* gene or p53 protein.

- *TP53* gene is a tumor suppressor gene
- Also known as *TP53*, phosphoprotein p53 antigen, NY-CO-13, transformation-related protein (*TRP53*).
- Mutation of p53 gene → produces more than 50% of all human tumors.
- Most of the mutations of *p53* gene are missense type.

Location
Chromosome 17(17p13.1)

Role of TP53 gene

TP53 gene encodes for a *p53 protein* (transcription factor) in response to DNA damage. This p53 arrests cell cycle or induces apoptosis. Hence, mutated *TP53* gene cannot control cell proliferation of cancerous cells **(Flow chart 1)**.

Note:
- Benzopyrene (in cigarette smoke), aflatoxin B1 produces TP53 mutation.
- *TP53* gene arrests the cell cycle in G1/S phase.
- Presence of *TP53* gene mutation indicates a more invasive form of breast and colon cancers.
- Gene therapy with normal *TP53* gene may treat cancers.

Regulation of Cell Growth (Fig. 1)

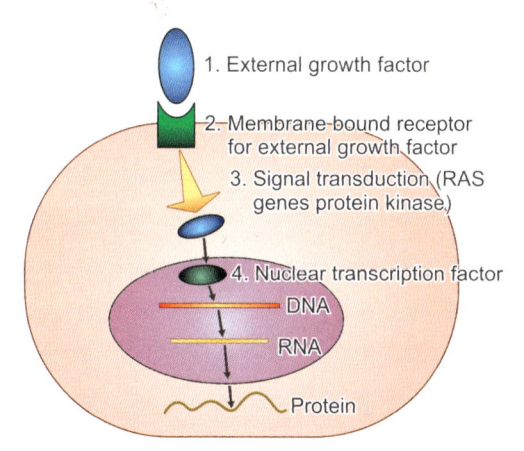

Fig. 1: Regulation of cell growth. It is controlled at four levels: (1) By external growth factors (such as steroid hormone, epidermal growth factors); (2) By binding of growth factors with its receptors; (3) By activation of signal transduction molecules (such as protein kinases) that reacts with nuclear transcription factors; (4) By nuclear transcription factors that regulate cell growth through regulating regions of DNA.

Flow chart 1: Role of *TP53* gene

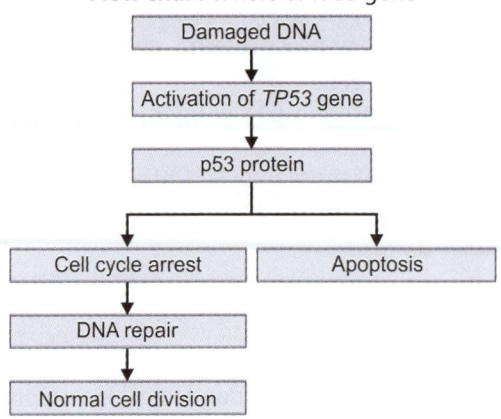

Damaged DNA → Activation of *TP53* gene → p53 protein → Cell cycle arrest → DNA repair → Normal cell division

p53 protein → Apoptosis

Box 1: Li-Fraumeni syndrome

Li-Fraumeni syndrome
- Li-Fraumeni syndrome is an autosomal dominant disorder that involves mutation of *p53* gene in germline cells.
- It is characterised by early onset of cancers such as sarcomas, breast, leukemia and adrenal gland cancers (hence, known as SBLA syndrome).

Some Interesting Facts

- Retroviruses convert an RNA into a DNA with the help of reverse transcriptase.
- Some viruses produce human tumors, such viruses are known as oncoviruses.
- Examples of oncovirus are given in **Table 3**.

Table 3: Oncovirus

Virus	Associated tumor
Hepatitis B virus	Hepatocellular carcinoma
Hepatitis C virus	Hepatocellular carcinoma
Human T-lymphotropic virus (HTLV)	Adult T-cell leukemia
Human papilloma virus (HPV)	Warts, precancerous lesions, genital cancers
Kaposi's sarcoma-associated herpes virus-8	Kaposi's sarcoma
Epstein-Barr virus (EBV)	Burkitt's lymphoma, Hodgkin's lymphoma, nasopharyngeal carcinoma

CHAPTER 11

Prenatal Diagnosis

CLINICAL CASE

A pregnant woman visited the antenatal clinic for a checkup at 12 weeks of pregnancy. She and her husband both are suffering from β-thalassemia minor, and now they want to know that whether their baby will be normal or a thalassemia patient? What test will you suggest for the diagnosis?

INTRODUCTION

- Prenatal diagnosis or screening is the way of prenatal care to determine the wellbeing state of the foetus and to detect the presence of any congenital anomaly or disorder in the foetus.
- It also includes the steps towards the decision for the treatment, counseling of the couple and termination of pregnancy.
- **Box 1** includes the indications for the prenatal diagnosis.

Q. Enlist the indications for prenatal diagnosis.

Box 1: Indications for prenatal diagnosis *(Viva)*

Indications for prenatal diagnosis
- Maternal age more than 35 years
- Family history of congenital anomalies
- History of the previous child with birth defects or chromosomal abnormalities, e.g. X-linked diseases
- Known genetic disorders of parents, e.g. thalassemia
- History of diabetes mellitus, phenylketonuria
- Infection during pregnancy
- Exposure to contraindicated drugs or known teratogen
- Ultrasound examination showing foetal abnormalities
- Overanxious mothers
- Population at increased risk of genetic diseases, e.g. Indians for sickle cell anemia.

METHODS OF PRENATAL DIAGNOSIS

Q. Enlist methods of prenatal diagnosis.
- There are many methods available for prenatal diagnosis.
- These methods can be classified as noninvasive, minimally invasive and invasive methods **(Table 1)**.

Table 1: Methods of prenatal diagnosis *(Viva)*

Type	Test
Noninvasive methods	• Prenatal checkup of mother (external examination) • Ultrasound examination • Foetal Doppler for heart sound

Minimally invasive methods	• Maternal blood screening for
	  – Circulating foetal cells
	  – Circulating free foetal DNA
	  – β-hCG (human chorionic gonadotropin)
	  – Alpha-foetoprotein (AFP)
	  – Pregnancy-associated plasma protein A
	  – Estriol
	  – Inhibin A
	• Transcervical retrieval of trophoblast cells (cervical mucous, swabbing or lavage)

Invasive methods	• Amniocentesis • Chorionic villus sampling • Foetoscopy • Foetal blood sampling

- In prenatal examination, selection of the method depends on the finding of clinical history and examination of the patient. If possible, invasive procedures should be avoided as they increase the chance of abortions.

Ultrasonography in Prenatal Diagnosis

Q. Enlist uses of ultrasonography in prenatal diagnosis.
- Ultrasonography can be used from the 4th week of the pregnancy till the delivery of the baby.
- It is useful for the following purposes:
    - Confirmation of the pregnancy
    - Confirmation of the multiple pregnancies
    - Localisation of the placenta and position of the foetus
    - Determination of gestational age
    - To monitor the foetal growth
    - To detect developmental defects.
- Ultrasonography is useful for detection of the following developmental disorders:
    - Hydramnios
    - Oligohydramnios

Box 2: Nuchal translucency, 3D and 4D ultrasound scan

Nuchal translucency

Q. Write a short note on Nuchal translucency.
- Nuchal translucency is the accumulation of fluid under the skin at the back of baby's neck.
- Nuchal translucency is measured using ultrasonography between 11 and 13 weeks of gestational age *(Viva)*

3D and 4D Ultrasound Scan
- For detection of foetal anomalies, still picture of the baby can be generated by 3D ultrasound scan while 4D ultrasound scan can produce a moving 3D image of the foetus.

- Anencephaly
- Hydrocephalus
- Spina bifida
- Polycystic kidney
- Limb deformities and so on.

Amniocentesis

Q. Write a short note on Amniocentesis.

- Amniocentesis is a prenatal diagnostic invasive procedure to collect amniotic fluid for analysis.
- Amniocentesis is first introduced by Fritz Friedrich Fuchs and Polv Riis in 1956.
- Ultrasound-guided amniocentesis was first reported by Jens Bang and Allen Northeved in 1972.

Suitable gestational age: 14–20 weeks. *(NEET)*

Procedure

- With the aid of the ultrasound guidance, amniotic sac is tapped to collect 20 mL of amniotic fluid **(Fig. 1)**.
- Fluid is centrifuged to isolate the amniotic cells (cells of the foetal origin). These cells are grown in the culture medium and examined for chromosomal abnormalities by karyotyping.
- Supernatant amniotic fluid is screened for the alpha-foetoprotein (AFP) and other biochemical parameters.

Risk

- The risk of miscarriage is low (1 in 300–500 cases).
- Other risks are an injury to the foetus, umbilical cord and placenta.

Screening Benefits

It is useful for detection of the following conditions
- Chromosomal disorders such as Down syndrome, Turner syndrome and so on.

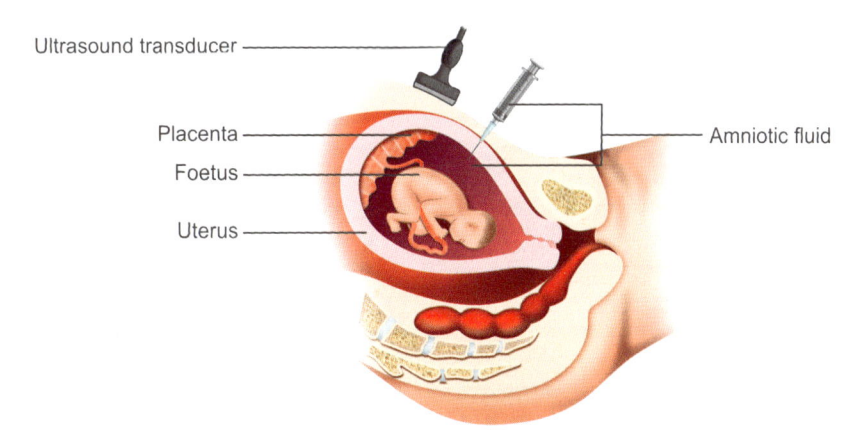

Fig. 1: Procedure for amniocentesis

- Genetic disorders such as cystic fibrosis, sickle cell anemia, Tay-Sachs disease.
- Neural tube defects such as spina bifida and anencephaly (increased AFP).

Limitations

Amniocentesis cannot detect structural birth defects such as cleft lip, cleft palate, heart malformations and so on. These can be detected by ultrasound.

Chorionic Villus Sampling

Q. Write a short note on chorionic villus sampling (CVS).

- Chorionic villus sampling is a prenatal test in that a sample of chorionic villi is aspirated from the placenta for testing.
- It was first time performed by Giuseppe Simon in 1983.

Suitable gestational age: 9–12 weeks. *(NEET)*

Procedure

- Chorionic villus sampling is done under ultrasound guidance.
- There are two approaches for CVS:
    1. *Transcervical approach*: In this method, a thin catheter is passed through the cervix into the uterus to collect cells of chorionic villi from the placenta **(Fig. 2)**.
    2. *Transabdominal approach*: In this method, a needle is inserted through the anterior abdominal wall into the placenta to remove chorionic villi cells.

Risk

- The risk of miscarriage is about 1–2%.
- Other risks include infection, amniotic fluid leakage and limb defects.

Merits

- Chorionic villus sampling can be done much earlier (in 9–12 weeks) as compared to amniocentesis (in 14–20 weeks). Hence, CVS can give early diagnostic results than amniocentesis.
- As CVS contains enough cells for genetic analysis, it provides rapid diagnosis than amniocentesis.

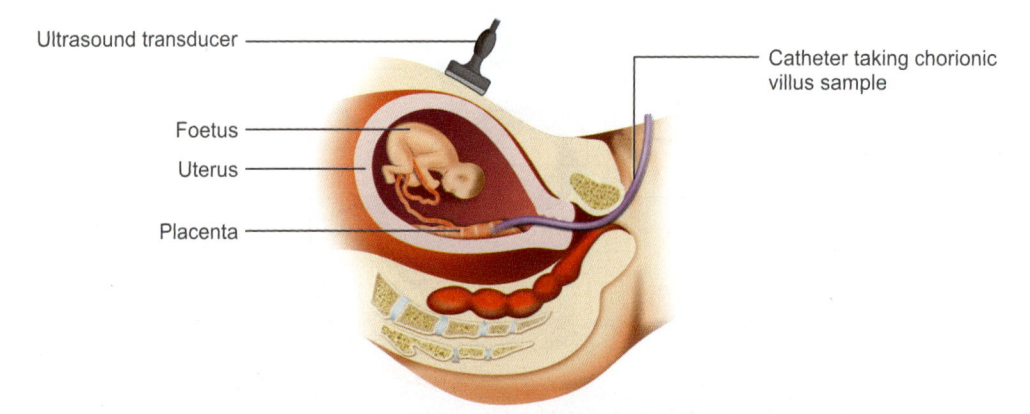

Fig. 2: Chorionic villus sampling—transcervical approach

- If the foetus is suffering from major genetic abnormality, due to early diagnosis by CVS, termination of the pregnancy may be planned safely in the first trimester itself.

Demerits and Limitations
- The risk of miscarriage is slightly higher in CVS than that in amniocentesis.
- In CVS, alpha-foetoprotein levels cannot be measured for the diagnosis of neural tube defects.
- Cells of CVS may show mosaicism (1%–2% cases), but it may not be there in the foetus.
- *Maternal cell contamination:* Placental blood may contain maternal cells.
- Chorionic villus sampling does not rule out structural abnormalities of the foetus.

Maternal Blood Screening Test

Q. Write a short note on maternal blood screening test.
- Maternal blood screening tests are performed for confirmation of pregnancy and to detect birth defects.
- Maternal blood screening is done for the detection of following substances:
    - Circulating foetal cells
    - Circulating free foetal DNA
    - β-hCG (human chorionic gonadotropin)
    - Alpha-foetoprotein
    - Pregnancy-associated plasma protein A
    - Estriol
    - Inhibin A.

Triple Screening Test

Q. Enlist the components of triple screening test.
- It is also known as multiple marker screening test, AFP plus, Kettering test or Bort's test. *(NEET)*
- It is a maternal blood screening test.
- There are three components of this test, mentioned as follows **(Table 2)**:
- *Purpose:* To screen the cases for chromosomal abnormalities such as Down syndrome and neural tube defects such as spina bifida.
- *Suitable gestational age:* 15–18 weeks.
- Sensitivity: 70% for detection of Down syndrome.

Quad Screening Test

Q. Enlist the components of quad screening test.
- It is a maternal blood screening test.
- *Purpose and time:* Similar to triple test.
- There are four components of quad screening test, mentioned as follows **(Table 3)**:

Table 2: Components of triple screening test

Component *(NEET)*	Source
1. Maternal serum alpha-foetoprotein (AFP)	Foetus
2. Human chorionic gonadotropin (hCG)	Placenta
3. Unconjugated estriol (uE)	Foetus and placenta

Table 3: Components of quad screening test

Component (NEET)	Source
1. Maternal serum alpha-foetoprotein (AFP)	Foetus
2. Human chorionic gonadotropin (hCG)	Placenta
3. Unconjugated estriol (uE)	Foetus and placenta
4. Inhibin A (IA)	Placenta and ovary

Percutaneous Umbilical Blood Sampling

Q. Write a short note on percutaneous umbilical blood sampling.

- Percutaneous umbilical blood sampling (PUBS) is also known as foetal blood sampling, cordocentesis, or umbilical vein sampling.
- Under ultrasound guidance, PUBS was first performed by Daffos et al. in 1983.

Procedure
Under ultrasound guidance, a thin needle is inserted into the umbilical cord through the anterior abdominal wall of mother to collect foetal blood.

Purpose
- To detect chromosomal, genetic and hematological abnormalities rapidly. It gives results within 2–3 days.
- To rule out mosaicism that is detected by amniocentesis or CVS.

Risk
- The risk of miscarriage is about 1%–2% cases.
- Other risks include infection and blood loss from the site of the puncture.

Fetoscopy
- Fetoscopy is the endoscopic visualisation of the foetus, amniotic cavity, umbilical cord and the foetal surface of the placenta.
- Under ultrasound guidance, foetoscope is introduced through the abdominal wall of the mother.

Some Interesting Facts

Acid elution test or Kleihauer-Betke test (1957)

This test is performed to detect the amount of foetal hemoglobin transferred from the foetus to the maternal blood. *(NEET)*

Purpose
- To determine the required dose of Rho immunoglobulin that inhibits the formation of Rh antibodies in Rh-negative mother and to prevent Rh disease in future in the Rh-positive foetus.

Lecithin-sphingomyelin Ratio
- Lecithin-sphingomyelin ratio (L/S ratio) in amniotic fluid is performed to assess the foetal lung maturity **(Table 4)**. *(NEET)*

Contd...

Contd...

- Surfactant (contains lecithin, sphingomyelin and other glycoproteins) is required for viability of the newborn in the external environment.

Urine Strip Test for Confirmation of the Pregnancy
- It is most widely used test for confirmation of pregnancy. This test depends on detection of β-hCG in maternal urine.
- It gives positive results after 10 days of first missed period. β-hCG can be detected after implantation.

Preimplantation Genetic Diagnosis
- Preimplantation genetic diagnosis (PGD) has been used to detect genetic defects before the implantation in in-vitro fertilisation (IVF) cases.
- In PGD, one or two cells are collected from fertilised ovum and will be screened for genetic disorders.

Table 4: Lecithin-sphingomyelin ratio

Lecithin-sphingomyelin ratio (*NEET*)	Indicates risk of foetal respiratory distress syndrome
More than 2	No risk, matured foetal lung
1.5–2	Mild–moderate
Less than 1.5	High risk

Box 3: Amniotic fluid index

Amniotic fluid index

Q. Write a short note on amniotic fluid index.
- The amniotic fluid index represents the amount of amniotic fluid observed on ultrasonography. *(Viva)*
- Amniotic fluid index is expressed in centimeters **(Table 5)**.

Table 5: Amniotic fluid index

Amniotic fluid index (*NEET*)	Indicated quantity of amniotic fluid
8–11 cm	Normal
Less than 5 cm	Oligohydramnios
More than 20 cm	Polyhydramnios

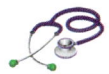

SOLUTION FOR CLINICAL CASE

β-thalassemia is an autosomal recessive disorder. In the present case, as both the parents are suffering from β-thalassemia minor, there is chance that the baby may be healthy, β-thalassemia minor or β-thalassemia major patient. CVS is a method of choice for prenatal diagnosis in this case as it can be done after 9 weeks any time and provides diagnosis earlier than amniocentesis. If the foetus is detected for β-thalassemia major, termination of pregnancy can be advised.

12
CHAPTER

Gene Therapy

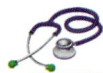

CLINICAL CASE

A diagnosed case of cystic fibrosis is referred for genetic counseling. The patient had a history of repeated lung infections. Now, the patient is asking to advise on the gene therapy. What would you like to suggest?

INTRODUCTION

Gene Therapy
Gene therapy is defined as an experimental technique that involves transfer, replacement or repair of the desired genetic material to cure or improve the clinical status of an individual. *(NEET)*

Vector
- Vectors are the vehicles used for delivering the foreign (desired) gene into the recipient cells from the patient.
- The vector can replicate autonomously and express the desired gene or manipulate the host cell genes.
- *Examples of vectors:* Plasmid, viruses, cosmids and artificial chromosomes.

HISTORY OF GENE THERAPY

Gene therapy is an emerging field. The reader must focus on gene therapy as it has rapidly developed since 2010.

Some of the milestones of gene therapy are listed in **Table 1**.

PRINCIPLES OF GENE THERAPY

Q. Write a short note on principles of gene therapy.

Gene therapy is meant to improve the clinical status of an individual with the following principles:
- Homologous recombination to replace the abnormal gene with a normal gene.

Table 1: Milestones of gene therapy

Year	Invention
1972	Article "Gene therapy for human genetic disease?" published. (Friedman and Robbin)
1984	Retrovirus vector system (Cepko et al.)
1990	National Institute of Health (NIH), United States of America (USA): First patient (Ashanti DeSilva) treated with gene therapy for severe combined immune deficiency by William Anderson.
1992	Gene therapy using hematopoietic stem cells (Claudio Bordignon)
1993	Cancer gene therapy for glioblastoma (brain tumor) (Trojan et al.)
2003	Liposome coated in polyethylene glycol to cross the blood-brain barrier (Ananthaswamy et al.)
2006	Gene therapy for the human myeloid system (Ott et al.)
2007	Gene therapy for inherited retinal disease (Leber's congenital amaurosis) (Maguire et al.)
2009	Nanotechnology-based gene therapy (Chisholm et al.)
2010	Gene therapy for color blindness in dogs (Komaromy et al.)
2010	Gene therapy for human beta-thalassemia major (Cavazza–Calvo et al.)
2011	Gene therapy for leukemia (Le Ford H)
2011	Plasmid DNA therapy for rat ischemic heart disease (Hahn et al.)
2014	NIH, USA: Clinical trial on the treatment of sickle cell anemia using umbilical cord stem cells
2017	Gene therapy for lung cystic fibrosis (Lee et al.)
2017	Gene therapy for non-Hodgkin lymphoma (Whipple et al.)
2017	Gene therapy for sickle cell anemia (Coghlan et al.)

- Selective reverse mutation to repair the abnormal gene and to make its function normal.
- Regulation of the gene expression (to increase or decrease gene function).
- Gene therapy is insertion, alteration or removal of a particular gene within the targeted cells of the tissue to treat or prevent the disease.

Some Interesting Facts

Which disease can be treated with gene therapy?
The candidate disease should have following characteristics:
- The disease should be inherited as recessive.
- It should have only one affected gene.
- Affected gene should be cloned.
- Precise regulation of gene expression should not be required.
- Affected cells should be accessible for the treatment.

APPROACHES FOR GENE THERAPY

For delivery of the gene, there are two approaches:
1. *Ex Vivo Approach*
    - *It involves:* Removal of host cells outside the body → introduction of normal gene or removal or repair of an abnormal gene in these cells → transplantation of these treated cells in the host body.

- *Example:* Addition of adenosine-deaminase gene using retroviruses in T-cells for patients with severe combined immunodeficiency patients.
2. *In Vivo Approach:*
   - It involves the direct introduction of a normal gene into targeted cells in the body of host.
   - *Example:* Transmission of cystic fibrosis transmembrane regulator *(CFTR)* gene using adenovirus (vector) for the treatment of cystic fibrosis.
   - Cystic fibrosis is an autosomal recessive disorder affecting *CFTR* gene located on chromosome 7. *CFTR* gene expresses chloride channel. As overexpression of this gene does not affect the cells, it is suitable for gene therapy. Cystic fibrosis affects lungs, pancreas and small intestine. In cystic fibrosis, there are increased mucous secretions and fatal lung infections.

Somatic Gene Therapy

- Somatic gene therapy is a technique that involves a change of the gene (insertion/replacement/repair of the gene) in somatic cells of the patient.
- As somatic gene therapy does not involve germ cells (sperms/ovum), the offspring of the treated individual will not have the corrected genes **(Table 2)**.

Germline Gene Therapy

- Germline gene therapy is a technique that involves a change of gene (insertion/replacement/repair of the gene) in germ cells (sperm/ovum).
- These changes will pass on to the offspring.
- Though germline gene therapy can cure future generations, due to ethical issues United States' government prohibited germline gene therapy and till date no human being is born with the treatment of germline gene therapy **(Table 2)**.

GENE TRANSFER TECHNIQUES

The gene transfer techniques are classified as follows **(Table 3)**:
- Physical methods
- Biological (vector-based) methods
- Chemical methods.

Table 2: Difference between somatic and germline gene therapy

Q. Write a short note on differences between somatic and germline gene therapy.

Somatic gene therapy	Germline gene therapy
1. Targeted genes are transferred to the somatic cells	1. Targeted genes are transferred to the germ cells
2. Later generations will not inherit transferred genes	2. Later generations will inherit transferred genes
3. Somatic gene therapy is currently in practice	3. Owing to ethical issues, germline gene therapy is not in practice
4. For example, the introduction of genes into lymphocytes, bone marrow cells, liver cells and so on	4. For example, the introduction of genes into the ovum and sperms

Table 3: Methods of gene transfer (*NEET*)

Physical methods	• Liposome-mediated gene transfer • Receptor-mediated endocytosis • Electroporation • Gene gun • Sonoporation
Biological methods	• Retroviruses • Adenoviruses • Herpes simplex virus • Adeno-associated viruses
Chemical methods	• Oligonucleotides • Polyplexes • Dendrimers

PHYSICAL METHODS FOR GENE DELIVERY

Q. Enlist the physical methods for gene transfer.

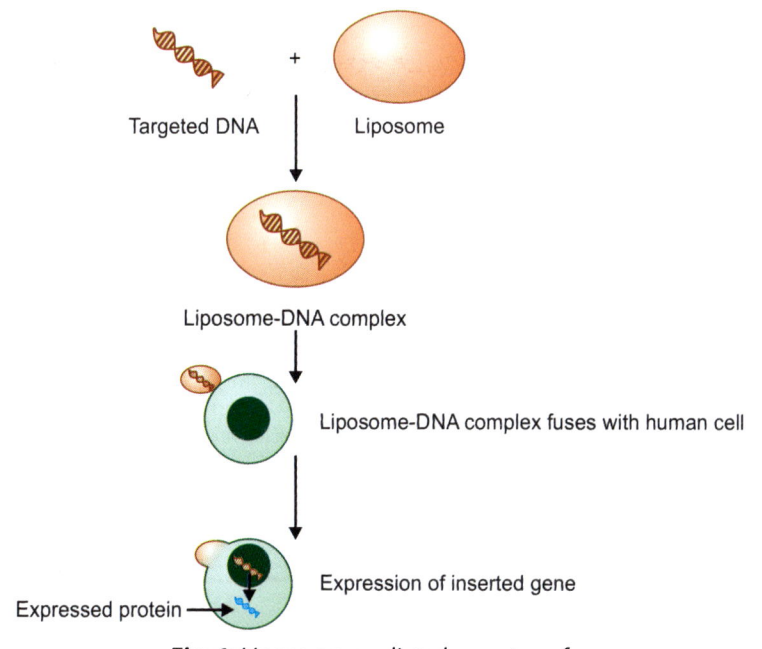

1. *Liposome-mediated gene transfer:*
   - Liposomes are spherical vesicles lined with the lipid bilayer.
   - Liposome-mediated gene transfer involves the introduction of a target gene in liposome → fusion of liposome with targeted human cells → expression of the gene in targeted cells **(Fig. 1)**.

Fig. 1: Liposome-mediated gene transfer

- *Advantage:* Large gene can be inserted.
- *Disadvantage:* Repeated treatment is required owing to transient gene expression.
2. *Receptor-mediated endocytosis:*
    - The ligand is a biomolecule that binds to the receptor molecules of the targeted cells to generate a biological response.
    - Receptor-mediated gene transfer involves targeted gene that is complexed with ligand molecules → injected into patient → gene-ligand complex bind with receptors of targeted cells → endocytosis of gene-ligand complex → expression of the gene **(Fig. 2)**.
    - *Advantage:* Viral vectors are not required. Hence, there is no risk of viral infections.
3. *Electroporation or Electropermiabilisation*
    - Electroporation involves the application of high voltage pulse to host cells → increase in permeability of these cells → entry of the targeted gene in cells → expression of the gene.
    - *Disadvantage:* High rate of cell death on electroporation **(Fig. 3)**.

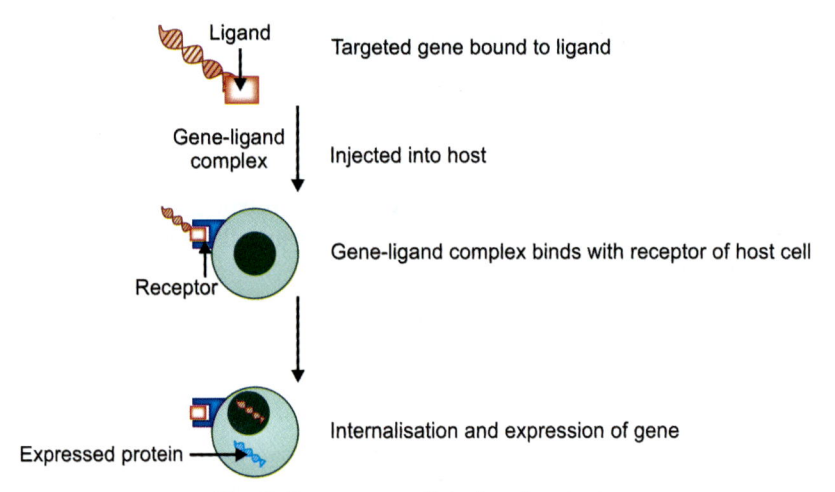

Fig. 2: Receptor-mediated endocytosis

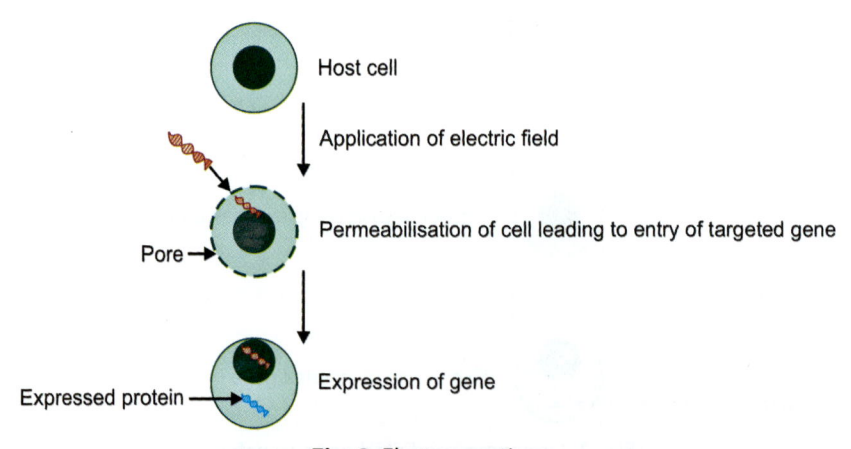

Fig. 3: Electroporation

4. *Gene gun or biolistic particle delivery system:*
   - Gene gun method involves coating of the targeted gene with gold particles → loading of gold-coated gene particles in gene gun → penetration of gene into the cells.
   - Gene gun is currently in use for delivering DNA vaccines.
5. *Sonoporation gene therapy:* Sonoporation includes: application of ultrasonic waves → formation of pores in the cell membrane → entry and expression of the targeted gene.

BIOLOGICAL (VECTOR-BASED) GENE TRANSFER

- In the biological gene delivery system, viruses are utilised as a vector for the mode of gene transfer to the host cells.
- For example, gene therapy for adenosine deaminase in cases of severe combined immunodeficiency (SCID) disease **(Fig. 4)**.
- In SCID, deoxyadenosine accumulates in T-cells → decreased immunity.

Viral Vectors for Gene Transfer

Q. Write a short note on viral vectors for gene transfer.

- Following viral vectors are commonly used for gene transfer:
  - Retroviruses
  - Adenoviruses
  - Herpes simplex virus
  - Adeno-associated viruses.
- All these viruses are deficient in encapsulation genes, and hence, these viruses can only produce proteins in the host cells but do not replicate.

Disadvantages of Viral Vectors

1. Induction of insertional mutagenesis
2. Provoke immune response
3. Limited gene insert size

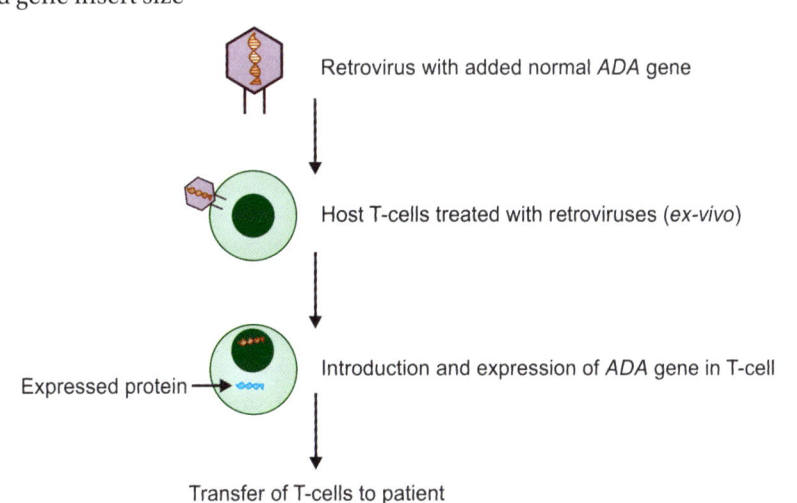

Retrovirus with added normal *ADA* gene

Host T-cells treated with retroviruses (*ex-vivo*)

Expressed protein → Introduction and expression of *ADA* gene in T-cell

Transfer of T-cells to patient

Fig. 4: Vector-based gene transfer
Abbreviation: ADA, adenosine deaminase

4. Unstable and transient gene expression
5. Lack of regulation of gene expression.

Advantages and disadvantages of commonly used viral vectors are given in **Table 4**.
Note: Blue-white screening is used to identify successful ligation of vector-bound gene in host cells. *(NEET)*

CHEMICAL METHODS OF GENE TRANSFERS

Q. Enlist chemical methods for gene transfer.

- *Oligonucleotides*: Used to inactivate genes that induce the disease.
- *Polyplexes*: These are complexes of polymers with DNA.
- *Dendrimers*: They are highly branched macromolecules that can be used as vector for gene transfer.

Table 4: Viral vectors and their advantages and disadvantages

Vector	Advantages	Disadvantages	Example
Retroviruses Utilises reverse transcriptase enzyme to form double-stranded DNA.	• Suitable for dividing cells such as bone marrow cells, liver cells, endothelial cells.	• Can deliver only small DNA (3–9 kb size). • Requires dividing cells, hence, not suitable for neurones. • No control on overexpression.	• Treatment of severe combined immunodeficiency.
Adenoviruses Double-stranded DNA viruses. Can infect respiratory tract, intestinal tract and eye.	• Useful for nondividing cells such as neurons. • Stable, easily purified. • Can carry large DNA (up to 30 kb size).	• Transient and unstable expression as they do not integrate host genome. • May produce infection. May cause malignancy.	• Delivery of the cystic fibrosis transmembrane regulator gene in cystic fibrosis.
Herpes simplex viruses	• Suitable for gene delivery in neurons, muscles, liver, pancreas and lung.	• Toxic effects on central nervous system. • Provokes immune responses. • Unstable and transient expression.	• Delivery of expressing "suicide" genes (thymidine kinase, cytosine deaminase, rat cytochrome P450) for the treatment of various tumors such as glioma, melanoma, breast, prostate, colon, ovarian and pancreatic cancer.
Adeno-associated viruses (parvovirus) Single-stranded DNA viruses.	• Can infect both dividing and nondividing cells. • Do not provoke an immune response or do not cause any disease. • Stable and sustained expression.	• Small packaging capacity (5 kb size).	• Human factor IX gene transfer in hemophilia B patients.

MEDICAL CONDITIONS TREATED WITH GENE THERAPY

Q. Enlist medical conditions treated with gene therapy.
- Adenosine deaminase deficiency
- Alpha-1 anti-trypsin deficiency
- Cystic fibrosis transmembrane regulatory enzyme deficiency in cystic fibrosis
- Low-density lipoprotein receptor replacement in familial hypercholesterolemia (Coronary artery disease)
- Complement C gene delivery in Fanconi's anemia
- Glucocerebrosidase replacement in Gaucher's disease
- Factor IX replacement in hemophilia B.
- Cytokine delivery in rheumatoid arthritis
- Insertion of tumor suppressor gene (*p53* gene) in Wilms' tumor
- Blocking expression of oncogenes in tumors
- Insertion of dystrophin gene in Duchenne muscular dystrophy.

DISADVANTAGES AND HURDLES IN GENE THERAPY

Q. Write a short note on disadvantages of gene therapy.
- Expensive
- Transient nature of therapy
- Provokes immune response
- Difficult to treat ***multigenic disorders*** such as Alzheimer's disease
- Wrong cells targeted by vector → insertion of gene at wrong place → ***insertional mutagenesis***
- Adverse effects of viral vectors
- Uncontrolled gene expression.

SOLUTION FOR CLINICAL CASE

The counselor should convey to the patient following facts:
- Cystic fibrosis is an autosomal recessive disorder involving *CFTR* gene.
- Abnormality of *CFTR* gene results in the production of an abnormal protein that causes mucous secretions and repeated lung infections.
- This disorder can be treated with gene therapy.
- Normal *CFTR* gene can be delivered to lung epithelia by using a simple inhaler loaded with adenovirus (vector) containing *CFTR* gene.

13
CHAPTER

Stem Cell Therapy

CLINICAL CASE

A pregnant woman visited OPD for an antenatal checkup. She had seen an advertisement regarding the umbilical cord blood preservation. Now, she wants to know about the utility and harm for the baby, charges and available facilities for the umbilical cord blood preservation. How will you respond to her inquiry?

INTRODUCTION

Stem Cells
Stem cells are undifferentiated cells that can self-replicate to form mature type of cells. These differentiated mature cells give rise to organs and tissues.

Differentiation
Differentiation is the process of development towards the complexity of cell organisation that results in specific cell formation.

Stem Cell Therapy
Principle: In the stem cell therapy, diseased organ cells are replaced with the help of stem cells. Stem cell therapy is also known as regenerative medicine.

Properties of Stem Cells (Fig. 1)
Q. Enlist the properties of stem cells.
- Infinite self-replication
- Capacity of differentiation (Potency)

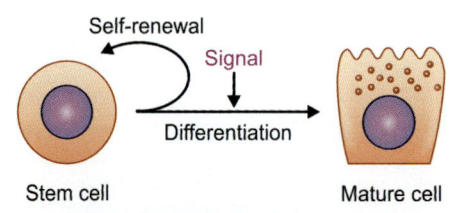

Fig. 1: Properties of stem cells

- Replaces old, damaged or dead cells
- Uncommitted until receiving a signal to form mature cell
- Immortal—provides endless supply of cells.

CLASSIFICATION OF STEM CELLS

Q. Write a short note on classification of stem cells.

Classification of Stem Cells According to Potency *(NEET)*

- *Totipotent:* Forms all differentiated cell types of an organism, for example, zygote, cells of morula.
- *Pluripotent:* Forms many differentiated cell types but not all, for example, inner cell mass of blastocyst can form all tissues except placenta and totipotent cells.
- *Multipotent:* Forms discrete cell types, for example, hematopoietic stem cells give rise to all kinds of blood cells.
- *Oligopotent/progenitor cells:* Differentiate into few cells, for example, lymphoid cells form T- and B-cells.
- *Unipotent:* Forms only one type of cell, for example, hepatocytes form only hepatocytes.

Differences between stem cells and progenitor cells are given in **Table 1**.

Types of Stem Cells According to the Source

- Stem cells are classified as:
    - Embryonic stem cells
    - Adult stem cells

Sources of Stem Cells

- Embryoblast
- Umbilical cord blood
- Teratoma cells
- Bone marrow cells, liver, epidermis, retina, skeletal muscle, intestine, dental pulp
- Aborted fetuses
- Placenta
- Amniotic fluid.

EMBRYONIC STEM CELLS

- Embryonic stem cells are obtained from inner cell mass (embryoblast) of the blastocyst **(Fig. 2)**.

Table 1: Differences between stem cells and progenitor cells

Stem cells	Progenitor cells
Unspecialised cells that can develop into a variety of specialised cell type.	Unspecialised cells that evolve into few types of specialised cell type.
Example: Hematopoietic stem cells can produce even neurons.	*Example:* Myeloid progenitor cell can produce only neutrophil and red blood cells.

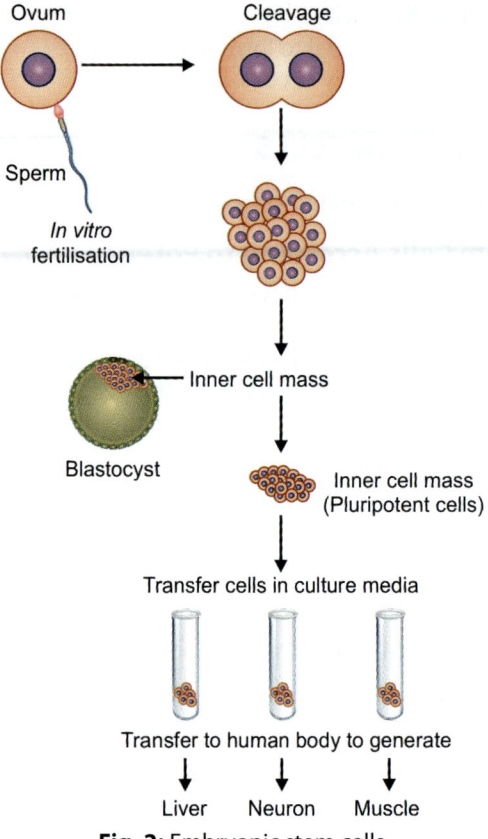

Fig. 2: Embryonic stem cells

- Cells formed immediately after fertilisation are totipotent → can form the whole embryo.
- However, cells from embryoblast are mainly pluripotent → form many tissues **(Box 1)**.

Advantages
- Multiply easily
- Can generate every cell type of adult body
- Immortal—can multiply indefinitely
- Can be manipulated easily for developing desired cells.

Disadvantages
- Difficult to differentiate uniformly
- Immunogenic—may be rejected after transplantation
- Tumorigenic—may form tumors
- Ethically not correct as it needs destruction of developing human life.

Box 1: Embryonic stem cell
- Till date, no therapeutic trial has been carried out for treating human disease with embryonic stem cells.

ADULT STEM CELLS

- These are also known as tissue-specific stem cells or somatic stem cells **(Fig. 3)**.
- These are specialised to some extent.
- These can produce some or all of the mature cell types of particular organ or tissue. Hence, adult stem cells are multipotent.
- For example, neuronal stem cell forms neurons, astrocytes and oligodendrocytes **(Box 2)**.

Advantages
- Nonimmunogenic
- Specialised to some extent
- Easy to procure
- Nontumorigenic
- No harm to the donor
- Ethically suitable.

Disadvantages
- Limited quantity
- Difficult to obtain
- Mortal—do not multiply infinitely like embryonic stem cells
- Mostly organ specific.

Differences between embryonic and adult stem cells are given in **Table 2**.

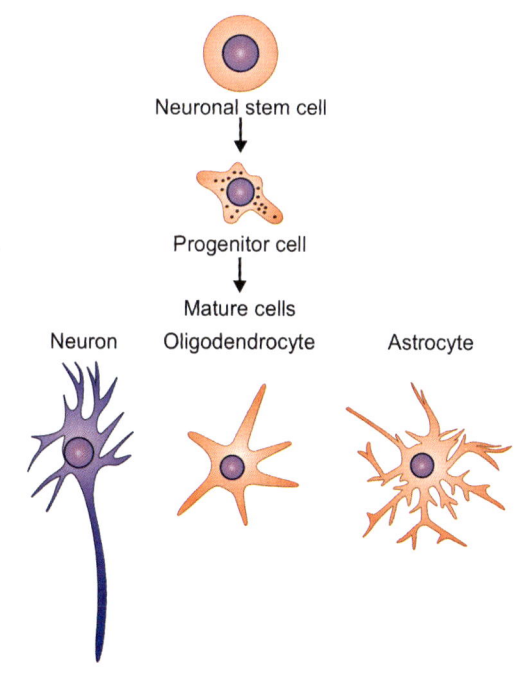

Fig. 3: Adult stem cells

Principles of Clinical Genetics

Box 2: Plasticity of stem cells
- It is the property of stem cell by which cells from one tissue can form completely different tissues.
- For example, hematopoietic stem cell forms heart muscles.

Table 2: Differences between embryonic and adult stem cells

Q. Write a short note on differences between embryonic and adult stem cells.

Embryonic stem cells	Adult stem cells
• Obtained from inner cell mass of the blastocyst	• Obtained from umbilical cord blood, bone marrow cells and so on
• Easy to get and multiply	• Difficult to obtain and multiply
• Immunogenic and tumorigenic	• Nonimmunogenic and nontumorigenic
• For collection, need destruction of developing human	• For collection, no major harm to the donor
• Multiply indefinitely	• Limited multiplication cycles
• Difficult to control differentiation	• Easy-to-control differentiation

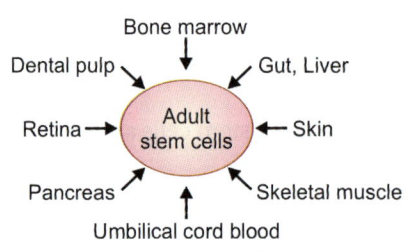

Fig. 4: Sources of adult stem cells

Sources of Adult Stem Cells (Fig. 4)
- Umbilical cord
- Placenta
- Stem cells in bone marrow, epithelia, dental pulp, brain, spinal cord and so on.

INDUCED PLURIPOTENT STEM CELLS

Q. Write a short note on induced pluripotent stem (iPS) cells.

   Mature (adult cell) can be converted into pluripotent stem cells as shown in **Figure 5**.
- Shinya Yamanaka (2006) first discovered iPS cells and got the Nobel Prize in 2012.
- Pluripotent genes include pivotal genes—*OCT4*, *SOX2*, *KLF4* and *c-MYC* gene.

APPLICATIONS OF STEM CELL THERAPY

Q. Write a short note on applications of stem cell therapy.

Potential applications of stem cell therapy are as follows:
- *Brain:* Stroke, traumatic brain injury, Alzheimer's disease, Parkinson's disease.
- Missing teeth

Stem Cell Therapy

```
Adult cell (mature cell)
        |
        | Transduce
        | stem cell associated genes
        | by viral vector
        ↓
Formation of
pleuripotent stem cells
        |
Autorenewal ↻
        |
        ↓
Differentiated into
different type of cells
```

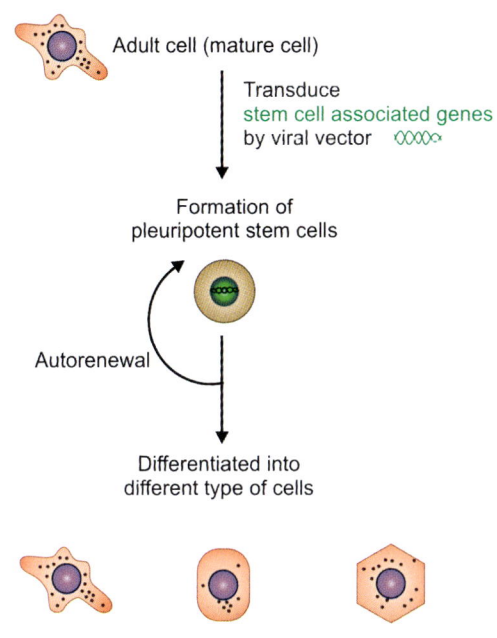

Fig. 5: Induced pluripotent stem cells

- Baldness
- Deafness
- Wound healing
- Bone marrow transplantation
- Myocardial infarction
- Muscular dystrophy
- Diabetes
- Spinal cord injury
- Genetic diseases
- Cancer patients treated with chemotherapy
- Aplastic anemia
- Severe combined immunodeficiency syndrome
- Thalassemia
- Cirrhosis of liver.

CORD BLOOD BANK

- Umbilical cord blood is good in stem cells content, hence can be used in the future for treatment of various diseases.
- A cord blood bank is a facility for the preservation of umbilical cord blood.

Protocol

Collection of 70–100 mL blood from umbilical cord within 10 minutes of birth
↓
Cryopreservation at −196°C

Cost: In India, it costs INR 40,000–60,000 for storage of cord blood. Cost mostly depends on the duration of storage (may be up to 75 years).

Cord Blood Banks in India

Cord blood banks in India are as follows:
- Cord Blood Bank (Government), Kolkata, India.
- LifeCell International Pvt Ltd, Chennai, Tamil Nadu.
- Jeevan Blood Bank and Research Centre, Chennai, Tamil Nadu.
- Cordlife Sciences (India) Pvt Ltd.
- Cryoviva Biotech Pvt Ltd, Gurgaon, Haryana.
- Reliance Life Sciences Pvt Ltd.
- NovaCord Fortis-Totipotent RX
- Babycell Regenerative Medical Services Pvt Ltd.
- Cryo-Save India Pvt Ltd.
- ReeLabs Pvt Ltd.
- StemCyte India Therpeutics Pvt Ltd.

Applications: Similar to the applications of stem cell therapy.

SOLUTION FOR CLINICAL CASE

- Umbilical cord blood is rich in hematopoietic or blood stem cells, and is currently used as an alternative to bone marrow transplantation.
- Umbilical cord blood can be collected noninvasively from the umbilical cord and placenta after birth, tested and stored frozen in tissue banks ready to use.
- In India, it costs INR 40,000–60,000 for storage of cord blood. Cost mostly depend on the duration of storage (may be up to 75 years). The expert personnel from cord blood bank attend the delivery, and they collect the sample.

14
CHAPTER

Genetic Counseling

CLINICAL CASE

A 30-year-old male patient was suspected of having chromosomal aberration and referred for the genetic counseling. The patient had a history of azoospermia. How will you deal the case at genetic counseling center?

INTRODUCTION

Q. Define genetic counseling.
- Genetic counseling is defined as the clinical counseling of a patient or his/her relative who is at the risk of a genetic disorder. *(Viva)*
- It is the counseling to make the individual aware regarding the genetic disease condition, risk of inheritance, and possible modalities for prevention and treatment.
- The person who performs genetic counseling is known as genetic counselor.

Aims or Goals of Genetic Counseling
The process of genetic counseling aims to:
- Provide information on genetic disorders.
- Diagnose the case and other family members who are at risk of genetic disorders.
- Assess the inheritance pattern of the disease.
- Provide medical and psychological support.
- Advise prevention of hereditary diseases.

Indicators for Genetic Counseling (Fig. 1)

Q. Enlist the indications for genetic counseling.
- Family history of genetic abnormality (A child with birth defect)
- Maternal age more than 35 years
- Foetal anomalies
- Consanguineous marriage
- Repeated miscarriages, infertility
- Teratogen exposure in pregnancy

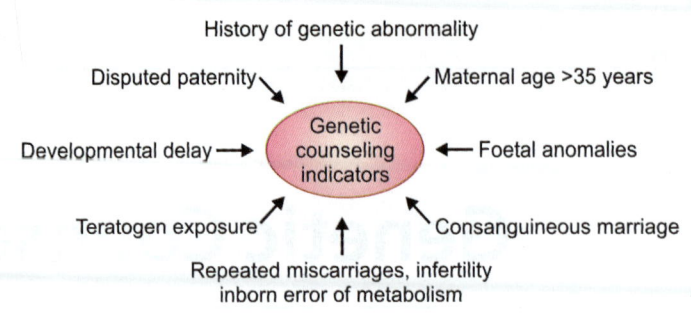

Fig. 1: Indications for genetic counseling

- Child adoption, developmental delay
- Case of disputed paternity
- Diagnosed case of genetic abnormality
- Suspected case of inborn errors of metabolism.

STEPS OF GENETIC COUNSELING

Q. Enlist the steps in genetic counseling.

The genetic counseling involves six phases as follows **(Flow chart 1)**:

Phase 1: Assessment
- This phase is related to acquire the knowledge regarding clinical history of the patient.
- It includes present and past history, family history, obstetric history (history of infertility, abortions, stillbirths) and consanguineous marriage.
- *Pedigree charting*: It helps to show the occurrence and appearance of a particular gene in family of an individual. It helps to assess the mode of inheritance.

Phase 2: Diagnosis
- The diagnosis of the disease can be done using following methods:
  - Phenotypic screening
  - Molecular or chromosomal analysis
  - Prenatal diagnosis
- The selection of method depends on the suspected disease and the status of patient (such as live-born baby, pregnant woman and so on).

Phase 3: Analysis
- It includes pedigree charting and estimation risk.
- Risk of the disease is estimated on the basis of mode of inheritance, pedigree charting and results of various diagnostic tests.

Phase 4: Communication
- The vital role of the counselor is to explain the information gained from analysis (Phase 3) to the patient.

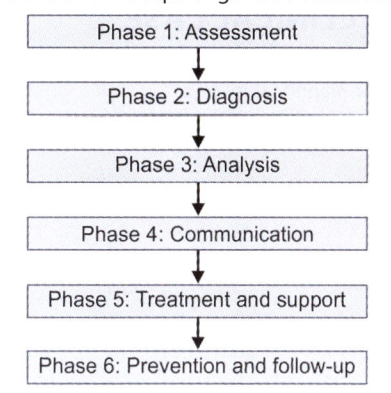

Flow chart 1: Steps of genetic counseling

```
Phase 1: Assessment
        ↓
Phase 2: Diagnosis
        ↓
Phase 3: Analysis
        ↓
Phase 4: Communication
        ↓
Phase 5: Treatment and support
        ↓
Phase 6: Prevention and follow-up
```

- Communication phase includes following factors:
  - Ethical transmission of information
  - Emotional support
  - Creation of right view in family members towards the disease
  - Education on disease (present and future status) and available treatments
  - Education on involved risk of inheritance
  - Advise on treatment, prevention and rehabilitation.

Phase 5: Treatment and Support
- Many genetic disorders do not have an available cure treatment. Only supportive measures are suggested to improve the status of an individual
- Advise on gene therapy and stem cell therapy is given depending on the disease condition, availability and economic conditions

Phase 6: Prevention and Follow-up
- Patient may be asked to visit multiple times for psychological counseling as per the need.
- If there is risk of transmission of the disease to the next generation, advise to be given on pregnancy (use of *in vitro* fertilisation, artificial insemination, prenatal diagnosis and so on) for prevention of disease transmission.

Box 1: Eugenics

Eugenics

Q. Write a short note on eugenics.

Definition
Eugenics is the science to improve the genetic quality of the human population by selective breeding.

Method for Eugenics
- Increasing rate of sexual reproduction in desired population (positive eugenics).
- Reducing rate of reproduction or sterilising affected (diseased) population (negative eugenics)

Current Practice
Eugenics is practiced in genetic counseling and prenatal diagnostics.

SOLUTION FOR CLINICAL CASE

The case should be dealt in a phasic manner as follows:
- Phase 1: Assessment
- Phase 2: Diagnosis
- Phase 3: Analysis
- Phase 4: Communication
- Phase 5: Treatment and support
- Phase 6: Prevention and follow-up.

ANNEXURES

ANNEXURE 1

Polymerase Chain Reaction

DEFINITION

- Polymerase chain reaction (PCR) is a molecular biology technique used to amplify a segment of DNA. It is *enzyme amplification* of the gene.
- Kary Mullis (1983) invented PCR.

REQUIRED MATERIAL

- Thermocycler is an equipment that increases or decreases temperature within a short span of time.
- PCR mixture **(Fig. 1)** *(NEET)*
  - DNA template to be amplified (isolated from a blood sample, bacteria and so on).
  - Buffer containing Mg^{2+} (required for DNA synthesis)
  - Taq DNA polymerase (heat-resistant enzyme) for the polymerisation of new DNA strand.
  - *Primers set:* Short chain of nucleotides that have complementary sequence, attaches to specific site on template DNA and begins DNA synthesis
  - *dNTP mix:* Containing mixture of deoxynucleoside triphosphates (building blocks of DNA)
- *Polymerase chain reaction tubes*: Small tubes used for amplification in a thermocycler.

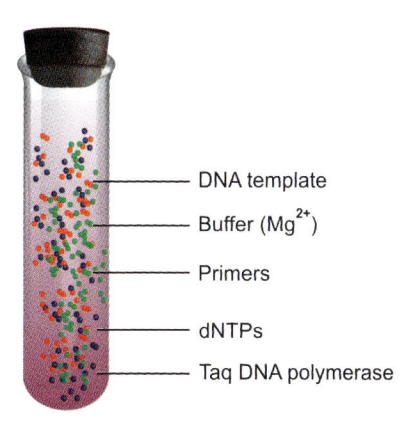

Fig. 1: Polymerase chain reaction master mix
Abbreviation: dNTPs, deoxynucleoside triphosphates

PROCEDURE

Polymerase chain reaction involves the following steps:

(a) *Denaturation*

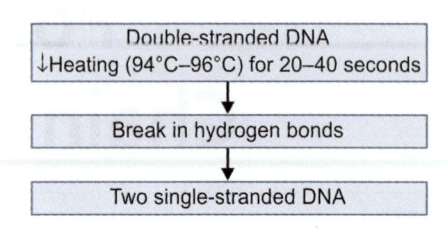

```
        Double-stranded DNA
    ↓Heating (94°C–96°C) for 20–40 seconds
                ↓
        Break in hydrogen bonds
                ↓
        Two single-stranded DNA
```

(b) *Annealing*

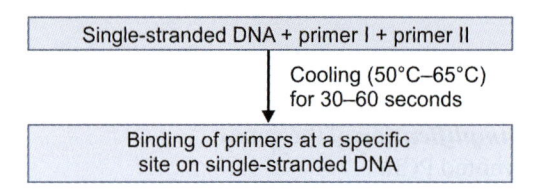

```
    Single-stranded DNA + primer I + primer II
                ↓ Cooling (50°C–65°C)
                  for 30–60 seconds
        Binding of primers at a specific
        site on single-stranded DNA
```

(c) *Elongation/Extension*

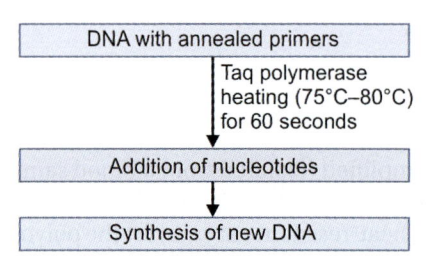

```
        DNA with annealed primers
                ↓ Taq polymerase
                  heating (75°C–80°C)
                  for 60 seconds
        Addition of nucleotides
                ↓
        Synthesis of new DNA
```

Here, the polymerase is used for DNA multiplication. *(NEET)*

(d) *Repetition of cycles*: 25–30 cycles are repeated to get sufficient copies of DNA segment.

(e) *Analysis of multiplied DNA segment* **(Fig. 2)**:

For analysis of multiplied DNA segment, one of the following methods can be used:
- *Agarose gel electrophoresis for estimation of size:* Agarose gel electrophoresis separates DNA fragments. *(NEET)*
- *Restriction enzyme mapping [restriction fragment length polymorphism (RFLP)]:* In RFLP, DNA separation is done by agarose gel electrophoresis. *(NEET)*
- Cloning and sequencing.

APPLICATIONS OF POLYMERASE CHAIN REACTION

- Selective DNA isolation
- Amplification and quantification of DNA
- Diagnostic applications
  - Detection of defective genes
  - Detection of infective agent DNA
  - Prenatal diagnosis
- Production of DNA vaccines
- Forensic applications—identification of criminals
- Paternity testing.

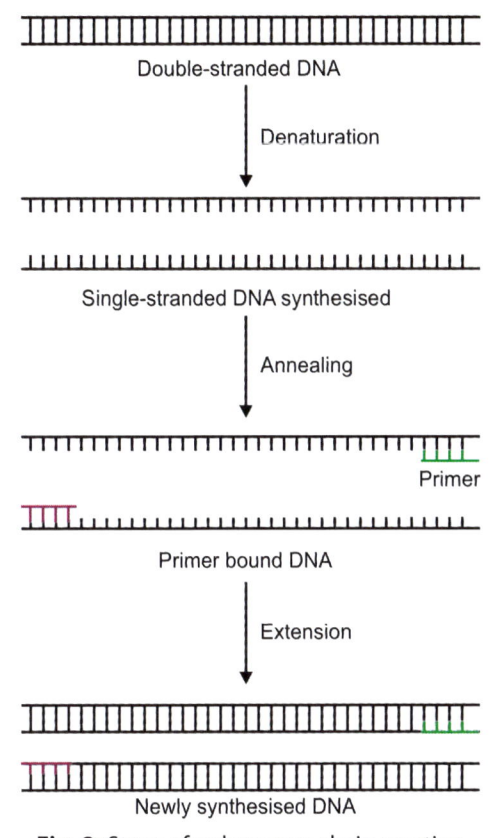

Fig. 2: Steps of polymerase chain reaction

Box 1: qPCT and rtPCR

Real-time polymerase chain reaction/qPCR
- It is a quantitative PCR.
- It helps to monitor amplification of targeted DNA molecule in real time and not only at the end as in conventional PCR.
- It is useful to quantify the targeted DNA present in the sample. For example, to find a viral load in patients' blood.

Reverse transcription PCR
- Reverse transcription PCR (rtPCR) is useful for detection of RNA.
- In rtPCR, enzyme reverse transcriptase is used for the conversion of messenger ribonucleic acid (mRNA) into complementary DNA (cDNA) prior to PCR **(Fig. 3)**.

Note:
- The number of DNA molecules formed from a single DNA molecule in PCR can be obtained by 2^n, here n is the number of cycles. *(NEET)*
- Nucleic acid absorbs light at 260 nm due to the presence of nitrogen bases. The absorbance is directly proportional to the concentration of DNA, hence, used for the quantification of the DNA. *(NEET)*

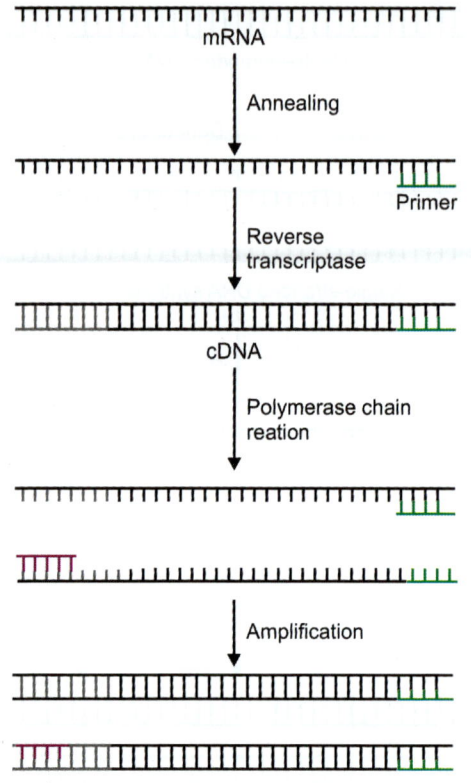

Fig. 3: Reverse transcription polymerase chain reaction
Abbreviations: mRNA, messenger RNA; cDNA, complementary DNA

- The mobility of the DNA molecules in agarose gel electrophoresis depends on the size of the fragment. *(NEET)*
- At physiological pH, DNA is negatively charged due to the presence of phosphate groups (PO_4^{-3}). *(NEET)*
- Gene can be amplified by PCR, ligase chain reaction, nucleic acid sequential-based amplification (NASBA) or cloning methods. *(NEET)*

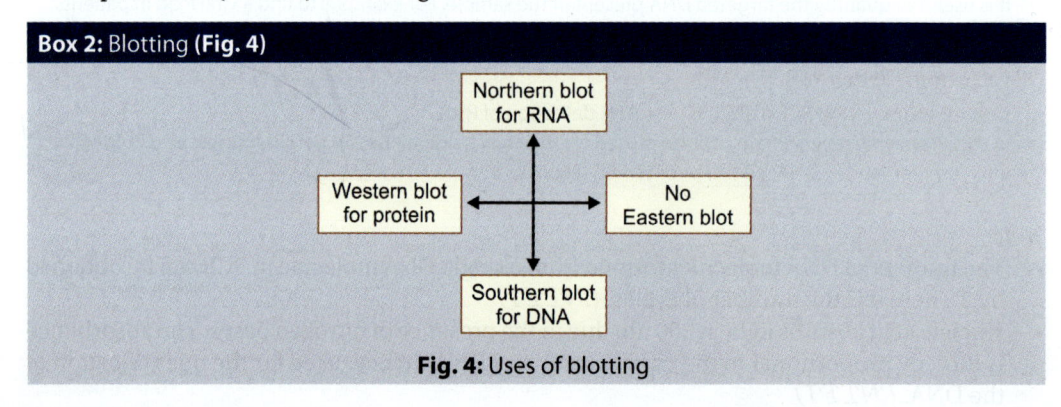

Box 2: Blotting (Fig. 4)

- Northern blot for RNA
- Western blot for protein
- No Eastern blot
- Southern blot for DNA

Fig. 4: Uses of blotting

Southern Blot Analysis
- Invented by Sir Edwin Southern (1975).
- Useful for identifying DNA sequence of interest.

Protocol

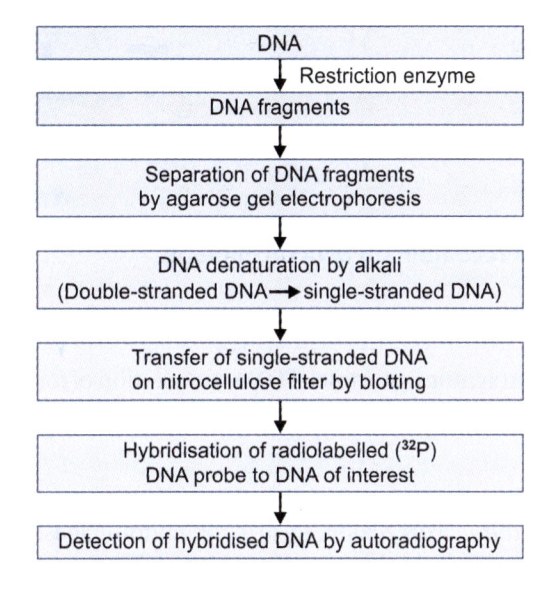

```
DNA
 ↓ Restriction enzyme
DNA fragments
 ↓
Separation of DNA fragments
by agarose gel electrophoresis
 ↓
DNA denaturation by alkali
(Double-stranded DNA → single-stranded DNA)
 ↓
Transfer of single-stranded DNA
on nitrocellulose filter by blotting
 ↓
Hybridisation of radiolabelled ($^{32}P$)
DNA probe to DNA of interest
 ↓
Detection of hybridised DNA by autoradiography
```

Northern Blotting
- Invented by Alwine (1979).
- Useful for identifying the presence of particular mRNA in a sample for diagnosis of disease and gene expression. *(NEET)*

Protocol

```
Extraction and purification of mRNA from cells
 ↓
Separation of mRNA by agrose gel electrophoresis
 ↓
Depurination for 5–10 minutes
 ↓
Blotting: Transfer RNA to amino
benzyl-p-methyl filter paper
 ↓
Hybridisation with DNA probes (radiolabelled)
 ↓
Autoradiography for detection of mRNA presence
(Note: Northern is a misnomer)
```

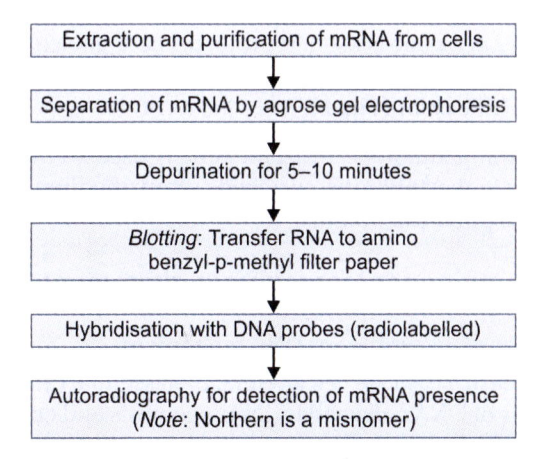

ANNEXURE 2

Recombinant DNA Technology

Q. Write a short note on recombinant DNA technology.

DEFINITION

It is a laboratory method to synthesise artificial DNA by insertion of foreign DNA into host DNA. As host cell replicate, the inserted DNA also gets multiplied.

PROCEDURE

- *Enzyme restriction endonuclease action:* Restriction endonuclease acts at a specific site on the vector (plasmid or bacteriophage) and selected DNA for multiplication. Restriction endonuclease cleaves double-stranded DNA. *(NEET)*
- *Enzyme DNA ligase activity:* DNA ligase seals sticky ends of selected segment of DNA with vector DNA.
- Transfection of host (*Escherichia coli* bacteria) with vector.
- Multiplication of host in culture to obtain multiple copies of selected DNA.
- *Expression of selected DNA-protein:* Host cells with selected DNA are allowed to multiply in enriched culture media to get desired protein.
- A plasmid containing *LacZ* gene is used. It produces beta-galactosidase that cleaves X-gal substances of culture media to produce a blue color. A plasmid with inserted DNA in *LacZ* gene cannot produce blue color **(Fig. 1)**.

Note: Vector DNA containing antibiotic-resistant gene is used so that only bacteria infected with such vector multiply in the antibiotic-enriched culture medium.
- DNA ligase forms phosphodiester bonds.

APPLICATIONS OF RECOMBINANT DNA TECHNOLOGY

Q. Enlist the applications of recombinant DNA technology.
- For diagnosis of a disease, i.e. sickle cell anemia, thalassemia, cystic fibrosis, and so on.
- For prenatal diagnosis of DNA collected by amniocentesis and chorionic villus sampling.
- For gene therapy of sickle cell anemia and other diseases.
- For the production of human proteins such as insulin and growth hormone. For this method, messenger ribonucleic acid (mRNA) isolated from beta-cells of the pancreas is used. *(NEET)*

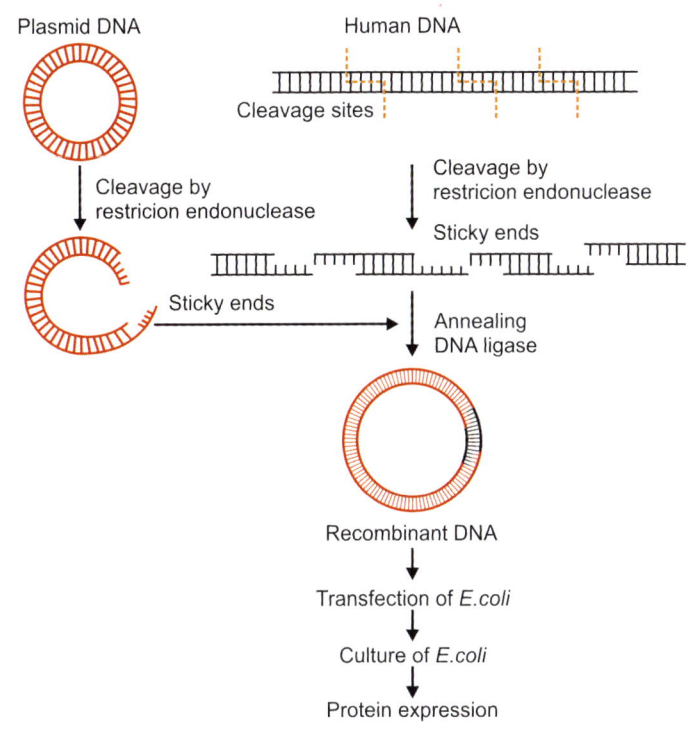

Fig. 1: Protocol for recombinant DNA technology

- For the production of DNA vaccines.
- For food industry, production of food additives.
- For agriculture, production of high-yielding breeds.

ANNEXURE 3

DNA Fingerprinting or Profiling

INTRODUCTION
- DNA fingerprinting is a molecular biology technique used to establish a link between the biological sample and the suspects in criminal investigation or is useful to establish paternity.
- Invented by Alec Jeffreys (1984).
- It is the method of restriction fragment length polymorphism (RFLP).

PROTOCOL

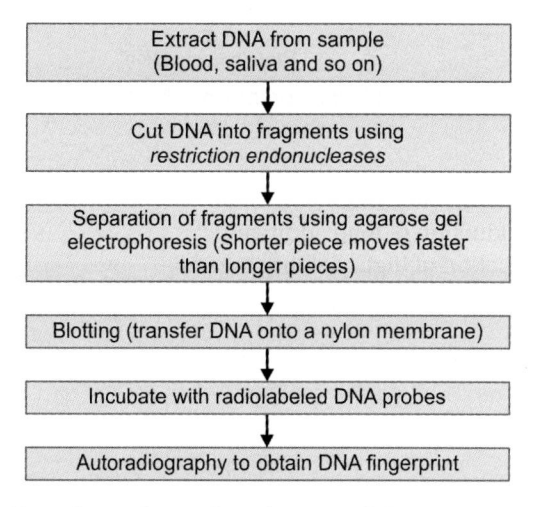

```
Extract DNA from sample
(Blood, saliva and so on)
           ↓
Cut DNA into fragments using
   restriction endonucleases
           ↓
Separation of fragments using agarose gel
electrophoresis (Shorter piece moves faster
           than longer pieces)
           ↓
Blotting (transfer DNA onto a nylon membrane)
           ↓
Incubate with radiolabeled DNA probes
           ↓
Autoradiography to obtain DNA fingerprint
```

- Currently, DNA profiling depends on identification of short tandem repeats (STRs) by using polymerase chain reaction.
- Splitting DNA or DNA from any nucleated cells can be used for DNA fingerprinting. *(NEET)*

ANNEXURE 4

Developmental Genetics

INTRODUCTION

- Developmental genetics involves the study of genes that control growth and development of an organism throughout the life cycle.
- Development is the conversion of a single cell into an organism.
- Development consists of:
  1. Cell division—to produce more cells.
  2. Cell differentiation—cell changes into a different type of cell.
  3. Morphogenesis—cells change their shape and structure to form a functioning organ.
- There are three genes that control the development of human. These are *HOX* genes, paired box genes and zinc finger genes.*(NEET)*

HOX (HOMEOBOX) GENES

Q. Write a short note on *HOX* genes.

- *HOX* genes control the development of an embryo by determining body axis (craniocaudal axis).
- *Functions of HOX genes*
  - Establish body plan during development.
  - Specify head and tail axis of embryo.
  - Specify positional identity of cells.
- Discovered by Bridges and Morgan (1923) in Drosophila.
- In human, homeobox family contains 235 functional genes.
- Four clusters of homeobox genes are present in human which are as follows:
  1. HOX A (*HOX* 1)—Chromosome 7p
  2. HOX B (*HOX* 2)—Chromosome 17p
  3. HOX C (*HOX* 3)—Chromosome 12p
  4. HOX D (*HOX* 4)—Chromosome 2q
- Some of the *HOX* gene subtypes are found to cause human diseases **(Table 1)**.

PAIRED BOX (PAX) GENE

- *PAX* genes code for tissue-specific transcription factors that control development of specific tissue.

- So far, nine *PAX* genes are identified (*PAX* 1–9).
- Some of their functions are listed in **Table 2**.

Table 1: Human *HOX* gene disorders

Gene	Disorder
HOXA1	Bosley-Salih-Alorainy syndrome, Athabascan brainstem dysgenesis syndrome
HOXA11	Radioulnar synostosis and thrombocytopenia
HOXA13	Hard-foot-genital syndrome, Guttmacher syndrome
HOXC13	Ectodermal dysplasia syndrome
HOXD10	Congenital—Marie-Tooth disease
HOXD13	Syndactyly, brachydactyly

For further reading, refer Quinonez SC, Innis JW. Human *HOX* gene disorders. Mol Genet Metab. 2014;111:4-15.

Table 2: *PAX* genes

Gene	Function
PAX1	Development of vertebrae and embryo segmentation
PAX2	Kidney and optic nerve. Mutation results in Renal–Coloboma syndrome
PAX3	Eye, ear, face development. Mutation → Waardenburg syndrome (premature greying of hairs, heterochromia of iris, deafness, patchy white skin pigmentation). Also found in alveolar rhabdomyosarcoma.
PAX4	Development of beta cells in pancreatic islets
PAX5	Neural, spermatogenesis development
PAX6	Development of eye and sensory organ
PAX7	Myogenesis Mutation → Aniridia and Peter's anomaly
PAX8	Thyroid gland development. Mutation → Thyroid anomalies and carcinoma
PAX9	Teeth development Mutation → Oligodontia

> **Box 1:** Zinc finger genes
>
> *Zinc finger genes*
> - Human genes contain more than 700 zinc finger genes.
> - These genes regulate transcription.
> - Mutation of zinc finger gene *GL13* (chromosome 7) causes *Greig cephalopolysyndactyly* (Cranial malformation with polysyndactyly) syndrome.

ANNEXURE 5

SRY Gene

Q. Write short note on *SRY* gene.
- *SRY* gene represents sex-determining region of Y chromosome.
- It codes for *testis determining factor* (TDF) protein.
- Testis determining factor is responsible for male sex determination.

ROLE OF *SRY* GENE

Role of *SRY* gene is explained in **Figure 1**.

MUTATION OF *SRY* GENE

- Mutation of *SRY* gene is associated Swyer syndrome (gonadal dysgenesis). *(NEET)*
- *SRY* gene is also related to the preponderance of dopamine-related diseases in males (such as schizophrenia, Parkinson disease).
- In XX male syndrome, one of the X chromosome, abnormally has *SRY* gene → genotypic female with phenotypic male. *(NEET)*

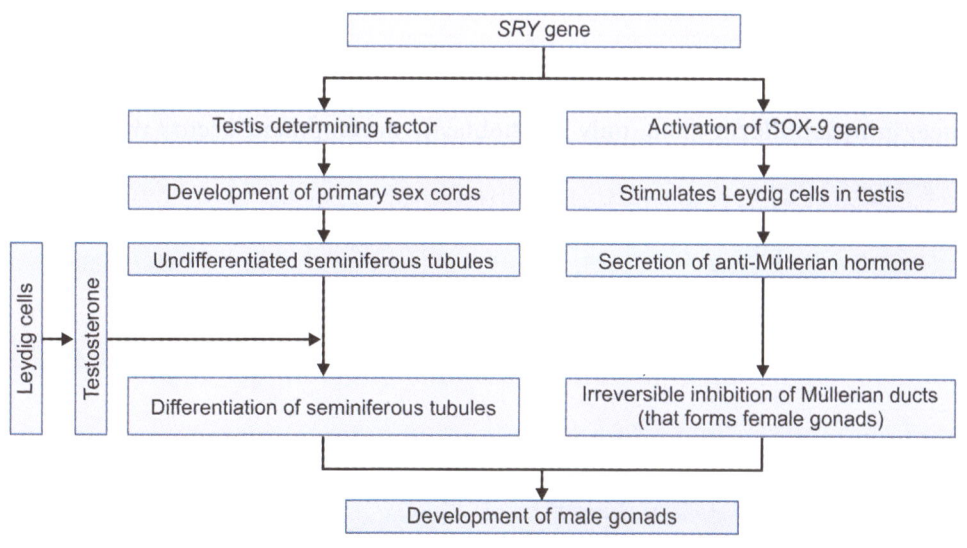

```
                        SRY gene
                   ┌───────┴───────┐
          Testis determining factor   Activation of SOX-9 gene
                   ↓                          ↓
          Development of primary sex cords   Stimulates Leydig cells in testis
                   ↓                          ↓
          Undifferentiated seminiferous tubules   Secretion of anti-Müllerian hormone
                   ↓                          ↓
          Differentiation of seminiferous tubules   Irreversible inhibition of Müllerian ducts
                                                    (that forms female gonads)
                   └───────┬───────┘
                   Development of male gonads
```

(Leydig cells → Testosterone → Differentiation of seminiferous tubules)

Fig. 1: Role of *SRY* gene

ANNEXURE 6

Hydatidiform Mole

Q. Write a short note on hydatidiform mole.
- It is also known as *molar pregnancy* in that nonviable fertilised ovum is implanted in the uterus.
- It is considered as a gestational trophoblastic disease that shows swollen *chorionic villi* resembling *clusters of grapes*.
- *Parthenogenesis:* If an egg develops into an individual without fertilisation, then condition is called parthenogenesis.

MODE OF FORMATION
- *Complete hydatidiform mole:* Ovum with lost nucleus gets fertilised with two sperms (thus, contains 46 chromosomes). Sometimes may get fertilised with single sperm that undergoes endoreduplication of the chromosome.
- *Partial hydatidiform mole:* Ovum fertilised with two sperm (triploidy) or more than two sperms (tetraploidy).

GENOMIC BASIS
- Maternal chromosomes help in normal embryo development.
- Paternal chromosomes are essential for trophoblast development.
- Hence, in hydatidiform mole, only trophoblastic development occurs through the contribution of paternal chromosomes.

ANNEXURE 7

Blood Group Genetics

MULTIPLE GENES

Q. Define multiple genes.
- *Alleles* are the genes controlling same *phenotypic character* and are located at the same gene *locus* of the homologous chromosomes.
- Mutated allele expresses different alternatives of same phenotypic character. Such genes are called multiple genes.
- An example of multiple genes is ABO blood group system and Rh factor.

ABO BLOOD GROUP SYSTEM

Q. Explain the genetic basis for ABO blood group system.
- Karl Landsteiner (1900) discovered ABO blood groups.
- Blood group depends on the presence of antigen on red blood cells (RBCs) and antibodies in plasma **(Table 1)**.

Inheritance of ABO Blood Groups
- Blood groups inherit by *multiple allelic* series.
- Human blood groups are inherited as gene *I* (Isohemagglutinin)
- There are three isoforms of the gene *I* as I^A, I^B and I^O.
- Gene I^A expresses antigen A, gene I^B expresses antigen B.
- Persons homozygous for two genes I^A and I^B express both antigen A and B.
- An individual with gene I^O expresses none of these antigens.
- I^A and I^B genes are dominant over I^O gene.

Table 1: Blood groups and corresponding antigen and antibodies

Blood group	Antigen	Antibody
A	A	Anti-B antibody
B	B	Anti-A antibody
AB	A and B	Nil
O	Nil	Anti-A and anti-B antibodies

Table 2: Blood groups with genotypes

Blood group	Genotype
A	I^AI^A, I^AI^O
B	I^AI^A, I^BI^O
AB	I^AI^B
O	I^OI^O

- Blood groups and corresponding genotypes are given in **Table 2**.
- When both the alleles are expressed together, for example, as in AB blood group, it is *called codominance*.

Rh FACTOR

Q. Write a short note on Rh blood group.
- Rh factor was discovered by Landsteiner and Wiener (1940).
- Rh factor depends on the expression of *D antigen*.
- D antigen is present in the rhesus monkey (hence, the individual with D antigen—Rh positive).
- D antigen gene is located on chromosome 1 (p36.13–p34.3) *(NEET)*.
- D antigen is a cell membrane bound glycoprotein of red blood cells (membrane transport protein).

Box 1: Erythroblastosis foetal is

Erythroblastosis foetal is:

Q. Write a short note on erythroblastosis foetalis.
- It is a haemolytic anaemia of newborn.

Pathology
- Rh negative woman and Rh positive man may have a Rh positive foetus.
- During pregnancy, few foetal RBCs with D antigen enter maternal circulation and produces anti-D antibodies in the mother.
- Anti-D antibodies cross foetoplacental circulation to induce destruction of foetal RBCs.
- This cause haemolytic jaundice and anaemia.
- In such mothers, first pregnancy may have a small quantity of antibodies.
- In a subsequent pregnancy, a large amount of anti-D antibodies is produced that may result in foetal death.
- Hence, during subsequent pregnancies, anti-D antibody levels in mother need to be monitored.
- After birth, the newborn baby may develop *kernicterus* due to deposition of the bile pigments in the brain and spinal cord. *(NEET)*

Prevention
- A shot of *anti-Rh antibodies within 72 hours of the delivery* of the baby should be given to Rh negative mother with Rh positive baby. *(NEET)*
- It will destroy baby's red blood cells that may have entered in the mother's blood before stimulating the mother's immune system for antibody production

- Person with D antigen—Rh positive
- Person without D antigen—Rh negative.

Box 2: Allele and codominance

Allele

A phenotypic character of an individual is controlled by a single pair of genes that occupies a specific position called gene locus on homologous chromosomes. These genes are called *alleles*.

Codominance

When both the alleles are expressed together, for example, as in AB blood group, it is called *codominance*.

ANNEXURE 8

Immunogenetics

IMMUNITY

Immunity is the ability of an organism to resist the infectious organism with the help of white blood cells.

Cellular Immunity

It is an immune response that is mediated by specific T-lymphocytes.

Humoral Immunity

It is an immune response that is mediated by the antibody (produced by B-lymphocytes).

ANTIGEN

The substance that induces an immune response is called antigen.

ANTIBODY

- The macromolecule (immunoglobulin) produced by B-lymphocyte that reacts with specific antigen, is called antibody.
- There are five types of immunoglobulins (Ig)—IgG, IgA, IgM, IgD, IgE.
- Immunoglobulins consist of:
    - Light polypeptide chains (L)
        - Kappa (κ): encoded by chromosome 2.
        - Lambda (λ): encoded by chromosome 22.
    - Heavy polypeptide chain (H): encoded by chromosome 14.
- Each chain has a constant region and a variable region.
- The variable region has an antigen-binding site that recognises specific antigen.
- The chromosome segment encoding heavy and light chain produces different messenger ribonucleic acids (mRNAs) due to splicing of genes and combinations of a large number of genes.

TYPES OF GRAFTS *(NEET)*

- Transplantation of tissue (grafting) may be required for treatment of diseases.

- The types of grafts are as follows:
  - *Autograft:* Graft tissue from one part to another part of the **same individual's** body.
  - *Isograft:* Graft of tissue between two genetically ***identical*** (monozygotic twins) individuals.
  - *Allograft:* Graft of tissue between genetically ***nonidentical*** individuals of ***same species***.
  - *Xenograft:* Graft of tissue between individuals of ***different species***.

MAJOR HISTOCOMPATIBILITY COMPLEX OR HUMAN LEUKOCYTE ANTIGEN SYSTEM

Q. Write a short note on human leukocyte antigen (HLA) system.

Q. Write a short note on major histocompatibility complex (MHC).

Histocompatibility

- The antigenic similarity between recipient and donor is called as *histocompatibility*.
- Histocompatibility should be assessed prior to transplantation.
- Major histocompatibility complex is a *cluster of genes* that are located at chromosome 6p21. *(NEET)*
- It plays an important role in discriminating self and nonself antigen-presenting structures.
- Major histocompatibility complex genes are codominantly expressed in each individual, thus monozygotic twins show MHC (codominantly expressed genes also include ABO blood group antigens).
- Major histocompatibility complex genes are the most polymorphic genes present in the genome. *(NEET)*
- These molecules are classified as class I, class II and class III.

Class I Major Histocompatibility Complex

- These are glycoproteins.
- These are expressed on all *nucleated cells*. *(NEET)*
- Present processed antigens to cytotoxic T-cells.
- Responsible for *graft rejection*. *(NEET)*
- According to the locus on the chromosome 6, MHC I genes are classified as *HLA-A, HLA-B, HLA-C*.

Class II Major Histocompatibility Complex

- Found only on *antigen-presenting cells* such as macrophages, B-cells, M-cells and dendritic cells.
- Presents processed antigen to helper T-cells.
- According to the locus on the chromosome 6, MHC II genes are classified as DP, DQ and DR.

Class III Major Histocompatibility Complex

- It encodes components of the complement system proteins, inflammatory cytokines, tumor necrosis factors, heat shock proteins and so on.
- Through these factors, MHC III genes play a vital role in modulating immunity.

HUMAN LEUKOCYTE ANTIGEN COMPLEX

- Genes coding for MHC molecules form HLA complex at chromosome 6.
- Human leukocyte antigen complex shows A, B and C locus for MHC I and D locus for MHC II (DR, DQ, DPA1, DPA2, DNA, DOB, DQB, DQB2 and DQA2).
- These loci show polymorphism. Hence, genetically nonidentical individuals are unlikely to have identical HLA phenotypes.
- Major histocompatibility complex genes are the most polymorphic genes present in the genome. *Haplotype* is combination of allelic forms of HLA molecules on one chromosome.
- Abnormality of the HLA genes is associated with some disease **(Table 1)**.

Table 1: Diseases and associated human leukocyte antigen (HLA) haplotype

Diseases with increased risk	Associated allele
Ankylosing spondylitis, postgonococcal arthritis, acute anterior uveitis	HLA-B27
21-hydroxylase deficiency	HLA-B47
Systemic lupus erythematosus	HLA-DR2, HLA-DR3
Autoimmune hepatitis, Primary Sjögren syndrome	HLA-DR3
Rheumatoid arthritis	HLA-DR4
Celiac disease	HLA-DQ2 and HLA-DQ8
Diabetes mellitus insulin independent	HLA-DR3, HLA-DR4

Box 1: HLA compatibility—bare lymphocyte syndrome

HLA compatibility
- In genetically different individuals, tissue graft compatibility and rejection of transplant depends on human leukocyte antigen (HLA) molecules.
- Initiation of the immune response is triggered by a recognition of antigen by T-cells.
- Inhaled, ingested or injected foreign or exogenous antigens are taken up by antigen-presenting cells, which express HLA class II molecules on their surface.
- *Bare lymphocyte syndrome:* It is produced by mutation in MHC class II genes. It results in reduced number and defective activation of CD4+ T cells. It is usually fatal. Treatment modality includes bone marrow transplantation or gene therapy.

ANNEXURE 9

Twins

DEFINITION
Twins are the two offsprings produced in a single pregnancy.

INCIDENCE
- One in 80–90 births.
- Incidence of twins is increasing owing to the development of assisted reproduction, such as, in vitro fertilisation, use of clomiphene citrate and so on.

TYPES OF TWINS
Twins are classified as monozygotic and dizygotic twins **(Fig. 1)**. The differences between monozygotic and dizygotic twins are given in **Table 1**.

Monozygotic Twins

Q. Write a short note on monozygotic twins.
- Fertilised ovum (zygote) divides into two embryos and forms monozygotic twins. Monozygotic twins have similar sex, external (phenotypic) characters and genetic composition.
- *Incidence:* 3 per 1000 deliveries.
- Types of monozygotic twins:
  - Monozygotic twins are classified as monozygotic dichorionic, monochorionic diamniotic, monochorionic monoamniotic twins **(Table 2)**.

Dizygotic Twins

Q. Write a short note on dizygotic twins.
- It is also known as nonidentical twins or dissimilar twins.
- It is the most common type of twins.
- They have different genetic constitution.
- They may differ in sex.
- Mechanism
  - Two separate ova are fertilised by two separate sperms and implanted in the same ovarian cycle, hence called biovular twins.
  - They are always dichorionic diamniotic as they have two placenta, chorionic sacs and amniotic sacs.

Principles of Clinical Genetics

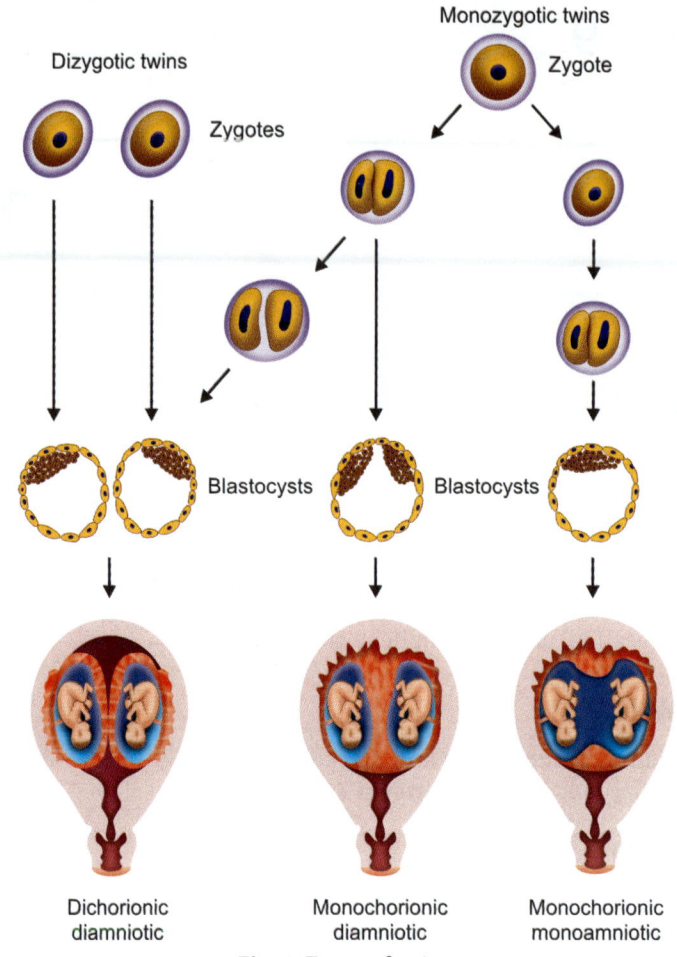

Fig. 1: Types of twins

Table 1: Differences between monozygotic and dizygotic twins

Q. Enlist the differences between monozygotic and dizygotic twins.

Monozygotic twins	Dizygotic twins
1. Product of one ovum.	1. Product of two ova.
2. An ovum is fertilised by a single sperm.	2. Separate ova are fertilised by separate sperms.
3. Have similar genetic composition.	3. Have separate genetic composition.
4. Have similar phenotypic characters.	4. May have different phenotypic characters.
5. Have same blood group.	5. May have different blood groups.
6. Rarely, they are dichorionic, diamniotic twins.	6. They are always dichorionic, monoamniotic twins.

Table 2: Types of monozygotic twins *(NEET)*

Type	Mechanism of formation	Placenta	Chorionic sacs	Amniotic cavities
Monozygotic dichorionic	Cell mass just after few cleavages separates into two masses	2	2	2
Monozygotic diamniotic (Most common type of monozygotic twins)	Inner cell mass of blastomere divides to form two separate embryos	1	1	2
Monochorionic monoamniotic	Bilaminar germ disc divides after amniotic cavity formation into two embryos	1	1	1

Box 1: Dichorionic diamniotic twins

Dichorionic diamniotic twins

All dizygotic and only few monozygotic twins are dichorionic diamniotic twins, that is, with two chorionic sacs and amniotic cavities.

Box 2: Conjoined twins

Conjoined twins
- These are monozygotic twins.
- *Mechanism of formation:* Inner cell mass of blastocyst fails to separate completely during twining that results in fused twins.
- These are monochorionic, monoamniotic twins.
- Types of conjoined twins: *(NEET)*
    - *Thoracopagus*: Fused chest wall up to umbilicus.
    - *Omphalopagus*: Fused lower abdomen.
    - *Parasitic twins*: Fused twins that have one smaller (rudimentary) twin.
    - *Craniopagus*: Fused twins at the base of the cranium.

ANNEXURE 10

Cloning

Q. Write a short note on cloning.
- *Clone:* Clone is defined as an organism that is produced by duplication of the genetically similar organism.
- Cloning is commonly employed for microorganisms.
- Monozygotic twins are the example of natural cloning in human beings. *(NEET)*
- Dolly sheep was the first cloned animal in the world (1997).

METHODS OF CLONING

Embryo Cloning
- It is artificial embryo twining.
- In this method, cells formed just after cleavages are separated from each other. At this stage, each of the separated cell has capability to form complete embryo. One of these cells is transferred to surrogated mother for further development.

Adult Cloning
- It is also called somatic cell nuclear transfer.
- In this method, fertilised ova are enucleated (removal of nucleus) and somatic cell nucleus is then transferred to enucleated fertilised ovum. Application of the electric current favors the fusion of the somatic cell with ovum and helps in transfer of the somatic nucleus. This ovum is transferred to the surrogate mother for further development. Dolly sheep is the example of an adult cloning.
- Due to ethical and legal issues, embryo and adult cloning in not permitted.

DNA Cloning
- This is the most commonly employed technique.
- It involves cloning of segment of DNA or messenger ribonucleic acid (mRNA) treated with restriction endonuclease.
- *Application of DNA cloning:*
  - Cloning of a segment of DNA is used in recombinant DNA technology.
  - Cloning of desired genes in a vector is useful for gene therapy.

Index

Page numbers followed by *f* refer to figure and *t* refer to table

A

ABO blood group 1, 111
  system 111
Achondroplasia 51, 53
Acid
  alpha-glucosidase 61
  elution test 78
Acyl-CoA dehydrogenase
    deficiency 61
Acyltransferase 56
Adenine 29-31
Adenosine 32
  deaminase 85
    deficiency 87
Adenoviruses 83, 85, 86
Adrenal hyperplasia, congenital 61
Adult stem cells 89, 91, 91*f*, 92, 92*t*
Adult T-cell leukemia 72
Agarose gel electrophoresis 100
Albinism 52, 61
Alkaptonuria 1*f*, 52, 61
Allele 47, 111, 113
Alpha-1 anti-trypsin deficiency 87
Alport syndrome 54
Alzheimer's disease 87, 92
Ambiguous external genitals 27
Amenorrhea 24
Amino acid metabolism, disease of 61
Amniocentesis 25, 74, 75, 75*f*
Amniotic cavities 119
Amniotic fluid 10, 79, 89
  index 79, 79*t*
Amplified fragment length polymorphism 40
Andersen disease 61
Anemia
  aplastic 93
  sideroblastic 58

Angelman syndrome 16
Aniridia 16, 108
Antibody 111, 111*t*, 114
Antigen 111, 111*t*, 114
Antioncogene 69
Aorta, coarctation of 24
Apoptosis 70
Arthritis
  postgonococcal 116
  rheumatoid 87, 116
Arylsulfatase B 62
Asthma 57
ATD angle 67, 67*f*
Athabascan brainstem dysgenesis syndrome 108
Autism 28
Autoimmune disorders 20
Autosomal dominant 49, 62
  disorders 51*t*
  inheritance 49, 51*f*, 53, 53*t*
Autosomal recessive 49, 62
  disease 63
  disorder 27, 52*t*
  inheritance 52, 52*f*, 53, 53*t*
Azoospermia 25

B

Baldness 93
Barr body 7, 7*f*, 9, 10
  number of 7, 7*t*
Barth syndrome 56
Becker's muscular dystrophy 56
Beta cells, development of 108
Biolistic particle delivery system 85
Blindness 56
Blood 10
  group 111, 111*t*, 112
    genetics 111
  system 54

Bone marrow 10, 92
  transplantation 93
Bort's test 77
Bosley-Salih-Alorainy syndrome 108
Bourneville-Pringle disease 51
Brachydactyly 108
Brain 92
Branched-chain-alpha-keto acid dehydrogenase 61
Breast cancer, hereditary 70
Bruton syndrome 56
Buccal smear 10, 25, 28
Burkitt's lymphoma 72

C

Cancer 69, 70*t*
  breast 70
  cervical 70
  colon 70
  colorectal 70
  genetics 2, 69
  genital 72
  ovarian 70
  type of 70
Carbamoyl phosphate synthetase 1 deficiency 61
Carbohydrate metabolism, disorders of 61
Carcinoma 108
  hepatocellular 72
  nasopharyngeal 72
  thyroid 70
Cat eye syndrome 21
Cataract 56
Celiac disease 116
Cell
  adhesion 70
  cycle regulation 70
  death 69

division 70, 107
growth, regulation of 71, 71f
proliferation 70
Centromere 4
    transverse division of 19
Chargaff's rule 30, 33
Chorionic sacs 119
Chorionic villus sampling 10, 25, 74, 76, 76f
Chromosomal
    aberration 1, 15, 20
        classification of 15
    disorders 48, 75
    duplication 20
Chromosome 3, 8, 12
    acrocentric 8
    artificial 80
    chemical composition of 7, 8
    length of 12
    morphology of 3
    philadelphia 19
    position of 12
    ring 16, 20, 20f
    shape of 12
    structural classification of 8
    structure of 4, 4f
    submetacentric 8
    telocentric 8
    types of 9f
Cleft lip 57
Cleft palate 26, 57
Clinodactyly 23
Cloning, methods of 120
Codominant inheritance 57
Coloboma syndrome 108
Color blindness 56
Copper metabolism 62
    disorders of 62
Cordocentesis 78
Cori's disease 61
Coronary artery disease 87
Cranial malformation 108
Cri-du-Chat syndrome 16, 17
Cubitus valgus 24
Cyclin-dependent kinase inhibitor 70
Cyclopia 27
Cystic fibrosis 52, 53, 82, 87
    transmembrane regulator 52
Cytochrome B gene 58
Cytosine 29-31

D

Deafness 93
Dendrimers 83, 86
Dental pulp 89, 92
Denver classification 12, 12t
Deoxynucleoside triphosphates 99
Deoxyribonucleic acid 1, 3, 8, 29, 58
Deoxyribose sugar 29
Depression 57
Dermal ridges, development of 64
Dermatoglyphics 64, 65, 68
    applications of 64
Diabetes 93
    mellitus 57
        insulin independent 116
Dichorionic diamniotic twins 119
DiGeorge syndrome 16
Diploid 21
    number 3
Distal interphalangeal creases 65
Distal palmar crease 65
Down syndrome 1, 21-23, 27, 64, 67, 75, 77
Duchenne muscular dystrophy 56, 87
Dystrophin 56
    gene, insertion of 87
Dystrophy, myotonic 51

E

Ectodermal dysplasia syndrome 108
Edward syndrome 21, 27
Embryoblast 89
Embryonic stem cell 89, 90, 90f
Endocytosis, receptor-mediated 83, 84
Enzyme 61, 104
    role of 62
Epidermal growth factor 70, 71f
Epidermis 89
Epilepsy 57
Epithelia 92
Epstein-Barr virus 72
Erythroblastosis foetalis 112
Escherichia coli bacteria 104
Estriol 74, 77
Estrogen 25

Euchromatin 5, 6, 6f, 6t
Exocrine pancreas dysfunction 58

F

Facial hemangioma 6
Fanconi's anemia 87
Fatty acid oxidation, disorders of 61
Favism 61
Fetoscopy 78
Feulgen reaction 13
Flat feet 28
Fluorescent *in situ* hybridisation 10, 13, 40
Focal dermal hypoplasia 54
Foetal
    anomalies 95
    blood 10
        sampling 74, 78
    cells 74, 77
    respiratory distress syndrome 79
Foetoscopy 74
Foetus 77, 78
    position of 74
Folic acid 20
Follicle-stimulating hormone 25
Forebrain development, failure of 27
Fragile leukemic cells 10
Fragile X
    mental retardation gene 54
    syndrome 28, 54

G

Galactose-1-phosphate uridyl transferase 62
Galactosemia 58
    classic 62
Galton classification 66
Gaucher's disease 62, 87
Gene 34
    bank 41
    cluster of 115
    complete set of 48
    delivery 83
    expression 37
    gun 83, 85
    jumping 1, 39
    mapping 39
        uses of 40

multiple 111
polydactyly 59
therapy 80, 81, 87
    disadvantages of 87
    milestones of 81*t*
    principles of 80
transfer 83, 85, 86
    chemical methods of 86
    methods of 83*t*
    techniques 82
Genetic
    counseling 95, 96, 96*f*
        steps of 96, 97
    diseases 93
    disorder 48
    mapping 40
    marker 40
Genitalia 27
Genotype 35, 43, 48, 112
Germ cell 82
    mutation 38, 39
Germline gene therapy 82, 82*t*
Gestational age, determination of 74
Giemsa stain 11
Glacial acetic acid 11
Glaucoma 57
Glioblastoma 70
Glucocerebrosidase 62
Glucose 6-phosphate dehydrogenase 61
    deficiency 56
Glutaric aciduria 61
Glutaryl-CoA dehydrogenase 61
Glycogen
    branching enzyme 61
    debranching enzyme 61
    storage disease 1, 61
Goltz syndrome 54
Gonadal dysgenesis 24, 109
Grafts, types of 114
Guanine 29-31
Guanosine triphosphate 32
Guthrie test 63
Guttmacher syndrome 108
Gynecomastia 25

H

Haploid 21
    number 3
Hard-foot-genital syndrome 108
Heart
    defects, congenital 57
    sound 74
Hemochromatosis 52
Hemoglobin beta gene 52
Hemophilia
    A 56
    B 56, 87
Hepatitis
    autoimmune 116
    B virus 72
    C virus 72
Hermaphroditism 27
Herpes simplex virus 83, 85, 86
Hers disease 61
Heterochromatin 5, 6, 6*f*, 6*t*, 13
    types of 5
Hexosaminidase A deficiency 62
High-resolution banding technique 13
Hodgkin's lymphoma 72
Homeobox genes 107
Homocystinuria 61
Homogentisate 1, 2-dioxygenase gene 52
Homogentisic acid oxidase 61
Homoploid 21
Horseshoe kidney 25
Human
    chorionic gonadotropin 77, 78
    genetics 1*t*
    genome project 40
        compilation of 1
    HOX gene disorders 108*t*
    leukocyte antigen 116*t*
        complex 116
        system 115
    papilloma virus 72
    T-lymphotropic virus 72
Hunter syndrome 62
Huntington chorea 51
Hurler syndrome 62
Hydatidiform mole 110
    complete 110
    partial 110
Hydroxylation 37
Hyperammonemia, cerebroatrophic 54
Hypercholesterolemia 51
    familial 87
Hyperplasia, adrenal 27
Hypertension 57
Hyperuricemia 56
Hypogonadism 25
Hypomelia 6
Hypotonia 23
Hypotrichosis 6
Hypoxanthine 29
    guanine phosphoribosyl transferase 56, 61

I

Iduronate sulfatase 62
Immunity 114
    cellular 114
    humoral 114
In vitro fertilisation 26, 79, 97
Indian National Gene Bank 41
Infertility 24, 25
Insemination, artificial 97
Intestine 89
Iris, heterochromia of 108
Isochromosome 16, 19, 20*f*

J

Juvenile gout 56, 61

K

Kaposi's sarcoma 72
Karyotype 7, 21, 26, 24
Kearns-Sayre syndrome 58, 61
Kettering test 77
Kidney 108
Kinetochore 4
Kleihauer-Betke test 78
Klinefelter syndrome 7, 21, 25, 26*f*, 67

L

Langer-Giedion syndrome 16
Leber's hereditary optic neuropathy 58
Lecithin-sphingomyelin ratio 79*t*
Lejeune's syndrome 17
Lesch-Nyhan syndrome 56, 61
Leukocoria 56
Li-Fraumeni syndrome 72, 72*f*
Limb deformities 75

Lipoprotein, low-density 51, 60, 87
Liposome 83
  mediated gene transfer 83, 83f
Liver 89
  cirrhosis of 93
  glycogen phosphorylase 61
Lymphedema over limbs 24
Lymphocyte syndrome 116
Lymphoid cells 89
Lyon hypothesis 7
Lysosomal
  enzyme acid
    sphingomyelinase 62
  storage disorder 52

M

Macroglossia 22
Major histocompatibility complex 115
Mandibulofacial dysostosis 51
Maple syrup urine disease 61
Marfan syndrome 51, 53
Marie-tooth disease 108
Maroteaux-Lamy syndrome 62
Maternal
  blood screening test 77
  cell contamination 77
  serum alpha-foetoprotein 77, 78
McArdle disease 61
Melanoma 70
  malignant 70
Mendel's laws 43, 46
  biological significance of 46
Mendelian disorders 48
Menkes disease 62
Mental
  disorders 63
  retardation 26, 28
Messenger ribonucleic acid 114, 120
Metacentric chromosome 8
Methylation 7
Micrognathia 22
Microphthalmia 27
Miller-Dieker syndrome 16
Mitochondrial
  functions, disorders of 61

inheritance 58, 58f
myopathy 58
Monosomy 21
Mosaicism 25
Motor neuron 1 gene, survival of 52
Mucopolysaccharides 62
Multigenic disorders 87
Multiple endocrine neoplasia 70
Muscle
  glycogen phosphorylase 61
  phosphofructokinase 61
Muscular dystrophy 93
Mutation 38, 108
  causes of 38
  chromosomal 39
  classification of 39t
Myeloid leukemia, chronic 19
Myocardial infarction 93
Myogenesis 108

N

Neonatal heel prick 63
Neural tube defects 57
Neurofibromatosis 51, 70
Neutral endopeptidase, phosphate-regulating 54
Neutrophil 7
Niemann-Pick disease 62
Norrie disease 56
Nuchal translucency 74
Nucleoside 29
Nucleotide excision repair 52
Nutritional deficiencies 20
Nystagmus, sensory 26

O

Okazaki fragments 41
Oligodontia 108
Oligonucleotides 83, 86
Oncogenes 69, 70t
Oncovirus 72, 72t
Optic nerve 108
Organic acid metabolism, disorders of 61
Oropharyngeal squamous cell carcinoma 70
Osteogenesis imperfecta 51
Osteoporosis 25

Osteosarcoma 70
Ovarian follicles 24
Ovary 78
Ovotesticular disorder 27
Ovum 3, 82

P

Pallister-Killian syndrome 21
Palmar creases 65, 66f
Parkinson's disease 92, 109
Patau syndrome 21, 26
Patchy white skin pigmentation 108
Pearson syndrome 58
Pedigree
  charting, advantages of 48
  symbols 49f, 50f
Pentose sugar 29
Percutaneous umbilical blood sampling 78
Peroxisomal function, disorders of 62
Peter's anomaly 27, 108
Phenotype 43, 48
Phenylalanine hydroxylase 52, 61, 62
  deficiency 1, 63
Phenylketonuria 1, 52, 53, 58, 61, 62
Phosphate molecule 29
Phytohemagglutinin 10
Placenta 77, 78, 89, 92
  localisation of 74
Plasma protein A, pregnancy-associated 74, 77
Pleiotropy 58
Polycystic kidney disease 51
Polydactyly 26
Polygenic inheritance 1, 56
Polymerase chain reaction 99, 99f, 101f
  applications of 100
Polyplexes 83, 86
Polyploid 21
Polyposis, familial adenomatous 70
Polysyndactyly syndrome 108
Pompe's disease 61
Porphobilinogen deaminase 61
Porphyria, acute intermittent 61

Prader-Willi syndrome 16
Pregnancy
    confirmation of 74, 79
    molar 110
Primers set 99
Progenitor cells 89, 89*t*
Progesterone 25
Protein synthesis 1
Pseudogenes 39
Pseudohermaphroditism 27
Pseudouracil 31
Puberty, delayed 25
Pyloric stenosis 57

Q

Quad screening test 77, 78*t*

R

Radioulnar synostosis 108
Real-time polymerase chain
    reaction 101
Red blood cells 111
Retina 89
Retinal ganglionic cell
    degeneration 58
Retinoblastoma 70
Retrotransposed pseudogenes 39
Retroviruses 83, 85, 86
Rett syndrome 54*f*
Reverse transcription polymerase
    chain reaction 102*f*
Ribonucleic acid 1, 31
Robert syndrome 6
Rocker bottom feet 26
Rubinstein-Taybi syndrome 16

S

Schiff's reagent 13
Schizophrenia 57, 109
Sclerosis, tuberous 51
Scoliosis 25
Severe combined
    immunodeficiency
    syndrome 85, 93
Sex chromatin, study of 10
Sickle cell
    anemia 52
    trait 47

Signal transduction genes 70
Simian crease 23, 23*f*, 26, 65
Single gene
    disorders 48
    inheritance 49
Single nucleotide polymorphism
    40
Sjögren syndrome 116
Skeletal muscle 89
Sly syndrome 62
Smith-Magenis syndrome 16
Somatic cell 82
    mutations 38
Somatic gene therapy 82
Sonoporation gene therapy 85
Sperm 3, 82
Spina bifida 75
Spinal cord 92, 112
    injury 93
Spinal muscular dystrophy 52
Spondylitis 57
    ankylosing 116
SRY gene 109
    mutation of 109
    role of 109, 109*f*
Stem cell 88, 89, 89*t*, 92, 93*f*
    classification of 89
    plasticity of 92
    pluripotent 92
    properties of 88, 88*f*
    sources of 89
    therapy 88
        applications of 92
    types of 89
Steroid
    metabolism, disorders of 61
    sulfatase enzyme gene 56
Strabismus 24
Stroke 92
Suwon crease 65
Syndactyly 108
Systemic lupus erythematosus
    116

T

Tafazzin gene 56*t*
Tandem mass spectrometry 63
Tarui's disease 61
Taurodontic teeth 25
Tay-Sachs disease 52, 62

Teratoma cells 89
Testicular biopsy 25
Testosterone 25
    low level of 25
    therapy 26
Tetrasomy 21
Thalassemia 93
Thenar crease 65
Thrombocytopenia 108
Thymine 29, 30
Thyroid gland development 108
Traumatic brain injury 92
Treacher Collins syndrome 51
Triple screening test 77, 77*t*
Triple X syndrome 21
Trisomy 21
    13 21
    18 21
    21 21
    8 21
    X 21
Trophoblast cells, transcervical
    retrieval of 74
Trypsin 12
Tumors 87
    suppressor gene 69, 70*t*, 87
    functions of 69
Turner syndrome 7, 9, 19, 21, 23,
    24*f*, 67, 75
Twins 117
    conjoined 119
    dizygotic 117, 118, 118*t*
    monoamniotic 119
    monochorionic 119
    monozygotic 117, 118, 118*t*,
        119, 119*t*, 120
    parasitic 119
    types of 117, 118*f*
Tyrosine 63
    deficiency 63
    kinase 56

U

Ullrich-Turner syndrome 23
Umbilical
    cord 89, 92
    vein sampling 78
Urea cycle, disorders of 61
Urine strip test 79
Uveitis, acute anterior 116

V

Ventricular septal defects 24
Viral vectors, disadvantages of 85
Viruses, adeno-associated 83, 85, 86
Vitamin D resistant rickets 54
von Gierke's disease 61

W

Waardenburg syndrome 51, 108
Warkany syndrome 21
Warts 72
Watson and Crick model 30
William syndrome 16

Wilms' tumor 6, 87
Wilson disease 62
Wolf-Hirschhorn syndrome 16
Wound healing 93
Wrist creases 65

X

Xanthine 29
Xeroderma pigmentosa 41, 52
X-linked
    agammaglobulinemia 56
    dominant 49
        inheritance 53, 54, 54t
    hypophosphatemia 54
    ichthyosis 56
    recessive 49, 62
        inheritance 54, 55f, 56t
XXX syndrome 7, 27

Y

Y chromosome, role of 1
Y-linked
    disorders 49
    inheritance 55

Z

Zinc finger genes 108
Zygote 117